# Ocean Waves Engineering

# International Series on Advances in Fluid Mechanics

**Objectives:**

The field of fluid mechanics is rich in exceptional researchers worldwide who have advanced the science and brought a greater technical understanding of the subject to their institutions, colleagues and students.

It is to bring such advances to the attention of the broad international community, that this book series has been established. Its aims are achieved by contributions to volumes from leading researchers by invitation only. This is backed by an illustrious Editorial Board for the Series who represent much of the active research in fluid mechanics worldwide.

Volumes in the series cover areas of current interest or active research and will include contributions by leaders in the field.

Topics for the series include: Bio-Fluid Mechanics, Biophysics and Chemical Physics, Boundary Element Methods for Fluids, Experimental & Theoretical Fluid Mechanics, Fluids with Solids in Suspension, Fluid-Structure Interaction, Geophysics, Groundwater Flow, Heat and Mass Transfer, Hydrodynamics, Hydronautics, Magnetohydrodynamics, Marine Engineering, Material Sciences, Meteorology, Ocean Engineering, Physical Oceanography, Potential Flow of Fluids, River and Lakes Hydrodynamics, Slow Viscous Fluids, Stratified Fluids, High Performance Computing in Fluid Mechanics, Tidal Dynamics, Viscous Fluids, and Wave Propagation and Scattering.

---



---

J. Noye
Department of Applied Mathematics
GPO Box 498
Adelaide
South Australia 5001

M. Ohkusu
Research Institute for Applied Mechanics
Kyushu University
Kasuga-koen 6, Kasuga
Fukuoka
Japan

W. Perrie
Bedford Institute of Oceanography
PO Box 1006
Dartmouth, Nova Scotia
Canada B2Y 4A2

H. Pina
Instituto Superior Tecnico
Av. Rovisco Pais
1096 Lisboa Codex
Portugal

H. Power
Wessex Institute of Technology
Ashurst Lodge, Ashurst
Southampton SO40 7AA
UK

D. Prandle
Proudman Oceanographic Laboratory
Bidston Observatory
Birkenhead
Merseyside L43 7RA
UK

S.K. Sarma
Department of Civil Engineering
Imperial College
Imperial College Road
London SW7 2BU
UK

H. Schmitt
Department of Fluid Mechanics Research
Institute of Theoretical Fluid Mechanics
Bunsenstrasse 10
D-3400 Göttingen
Germany

M.P. Singh
Department of Mathematics
Indian Institute of Technology
New Delhi
India

P. Skerget
Faculty of Technical Sciences
University of Maribor
PO Box 224, Smetanova 17
62000 Maribor
Slovenia

P.A. Tyvand
Department of Agricultural Engineering
Agricultural University of Norway
PO Box 5065
N-1432 AS-NLH
Norway

L.C. Wrobel
Wessex Institute of Technology
Ashurst Lodge, Ashurst
Southampton SO40 7AA
UK

M. Zamir
Department of Applied Mathematics
The University of Western Ontario
Western Science Centre
London, Ontario
Canada N6A 5B7

Advances in Fluid Mechanics
Volume 2
Series Editor: M. Rahman

# Ocean Waves Engineering

Editor:
M. Rahman
*Department of Applied Mathematics, Technical University of Nova Scotia, Halifax, Nova Scotia, Canada*

Computational Mechanics Publications
Southampton Boston

M. Rahman
Department of Applied Mathematics
Technical University of Nova Scotia
PO Box 1000
Halifax, Nova Scotia
CANADA B3J 2X4

Published by

**Computational Mechanics Publications**
Ashurst Lodge, Ashurst, Southampton, SO40 7AA, UK
Tel: 44 (0) 703 293223; Fax: 44 (0) 703 292853
E-mail: CMI@ib.rl.ac.uk

For USA, Canada and Mexico:

**Computational Mechanics Inc**
25 Bridge Street, Billerica, MA 01821, USA
Tel: (508) 667-5841; Fax: (508) 667-7582
E-mail: CMINA@com.netcom

**British Library Cataloguing-in-Publication Data**

A Catalogue record for this book is available
from the British Library

ISBN 1-85312-285-8 Computational Mechanics Publications, Southampton
ISBN 1-56252-209-4 Computational Mechanics Publications, Boston
ISSN 1353-808X Series

Library of Congress Catalog Card Number 94-70414



Printed and bound in Great Britain by Bell & Bain Ltd., Glasgow

# CONTENTS

# PREFACE

The field of ocean waves is over 150 years old and during this period of time a considerable number of books and research papers have been published. Perhaps the most outstanding is the *Hydrodynamics* book by Sir Horace Lamb.[1] Two other excellent books in this field are *Linear and Nonlinear Waves* by G.B. Whitham[2] and *Surface Waves* by J.V. Wehausen and E.V. Laitone[3]. Other notable books are R.L. Wiegel's[4] *Oceanographical Engineering* and G.B. Airy's *Tides and Waves.*[5] Meanwhile, however, there has been considerable development with respect to the analytical, numerical and experimental aspects of the subject, and this book contains seven state-of-the-art articles by leading authors in the field.

The first chapter has been written by P. McIver on 'Low-frequency asymptotics of hydrodynamic forces on fixed and floating structures'. This article is concerned with the time harmonic linearised theory of water waves and structures. By making the assumption that the frequency of the motion is sufficiently low, or equivalently that the waves are much larger than a typical body dimension, it is possible to derive simple analytic formulae for these wave forces on structures. These formulae give an understanding of the complex dependence of the force in relation to the geometry of the structure. The long wave limit of wave forces may also be used as a partial check on numerical calculations.

In this article the main methods for investigating low-frequency asymptotics are reviewed. Modifications for finite depth that do not appear to be widely known are given, together with well-known results for deep water. The main part of the article is concerned with how simple application of the method of matched asymptotic expansions may be used to obtain quick results for single element structures. The more complex issue of multi-element structures is also addressed using the method of matched asymptotic expansion.

Chapter 2 has been written by T. Matsui on 'Second-order sum-frequency wave loads and response of a tension leg platform'. This article considers a complete second-order diffraction theory which is formulated for computing second-order wave loads on floating bodies in regular waves. The theory is applied to predict the second-order sum-frequency response of a tension-leg platform. It is shown that the vertical plane motions and resulting tether forces of the tension leg platform include second-order resonant components which are comparable in magnitude with first-order wave frequency responses. Model tests have been performed to validate the theoretical predictions. Satisfactory agreement is observed between predicted and measured results, confirming the validity of the theory.

Chapter 3 is written by B.N. Mandal and S. Banerjea entitled 'Some radiation and diffraction problems in the linearised theory of water waves'. Assuming linearised theory of water waves diffraction and radiation problems are studied in this article. This class of problems reduce to boundary value problems (BVP) associated with

Laplace's equation for determination of the velocity potential. The BVPs satisfy a Churchill type boundary condition at one boundary (free surface) and Neumann type condition at other boundaries. A simplified perturbation technique is employed directly to the nearly circular cylinder. Some special BVPs arising here involve the integral equation approach originating from suitable application of Green's integral theorem or Havelock's expansion of water wave potentials.

The fourth chapter has been written by S. Roy Choudhury and L. Debnath on 'Painlevé analysis, similarity reductions and special solutions of non-linear evolution equations'. Painlevé analysis, stretching transformations, and Lie group analysis are employed to investigate the solutions of the Melnikov equations. The equations are found to possess only the conditional Painlevé property. Nontrivial similarity reduced solutions resulting from the use of stretching transformations are not reducible to elliptic functions or any of the Painlevé transcendents and, in general, may have movable essential singularities. Only trivial solution results from the use of Lie group techniques, and from using generalized expansions truncated at various levels. Painlevé expansions, truncated at various levels, are also used to derive new analytical solutions of the Zakharov-Kuznetsov equation. In particular, two families of shock-type solutions, different from known solitary wave solutions, are obtained. The first, three-parameter family solutions, consists of bidirectional travelling waves with variable phase speeds along both spatial coordinates. The second, two-parameter family of solutions, consists of arbitrary speeds rightward travelling waves along $x$ and bidirectional waves along $y$. The parameter may be chosen to model different sets of numerical or physical data.

Chapter 5 entitled 'Second-order loading on a pair of vertical cylinders' is written by M. Rahman and D.D. Bhatta. An estimation of the hydrodynamic forces to the second-order is presented in this article. Special attention is paid to derive the quadratic forces for a pair of bottom mounted, surface piercing circular cylinders in waves of arbitrary uniform depth. The theoretical results are based on the large spacing approximation and the method involves replacing scattered diverging waves by plane waves. The theoretical results are presented to illustrate the influence of the various waves and structural parameters on the hydrodynamic forces on each of the cylinders. The amplification factors for the first- and second-order drift forces are computed and presented in tabular form. The second-order forces on a single cylinder are also computed numerically and presented in graphical form for various wave parameters.

The sixth chapter has been written by J.S. Pawlowski and R.E. Baddour on 'Recent advances in numerical modelling of non-linear propagation of water waves and their interaction with impermeable bodies.' Numerical modelling of the non-linear propagation of water waves and their interaction with ships and floating or fixed structures is of much practical importance. The prediction and evaluation of operational performance and safety of those objects are increasingly performed by means of computer programs rather than physical experiments.

The inherent difficulty in the numerical modelling of the corresponding hydrodynamic boundary value problems exists in the presence of the unknown free surface on which non-linear and time dependent boundary conditions apply. Because of their level of difficulty those problems are at present dealt with primarily by using the assumption of potential flow.

In this chapter the authors present a discussion of the general problem of the non-linear interaction of ideal fluid free surface waves with fixed or moving impermeable bodies. The presentation is divided into three main sections. The first section explains and illustrates the use of a novel domain mapping technique, developed by the first author, which makes possible consistent formulation of perturbation type boundary value problems and their solutions, including consistent construction of non-linear wave radiation conditions. The second section describes the boundary element semi-discretization techniques which have been found by the authors to be suitable and reliable in obtaining numerical solutions to the non-linear boundary value problems in time domain. It also contains a description of the novel semi-discretization method by means of modal potentials. The last section describes specific applications of the presented theoretical and numerical modelling techniques. The applications comprise of novel non-linear time domain numerical models of ship interaction with heavy seas, including the use of weak scatter hypothesis, and a novel non-linear model of the generation of waves and wave resistance by a ship advancing in calm water. The last section uses the results obtained by the authors and their collaborators over the last five years. In the conclusion, promising directions for further development of the research are discussed.

Finally, Chapter 7 written by M. Aslam Noor covers 'Variational inequalities in physical oceanography'. In recent years, variational inequality theory has emerged as a fascinating and interesting branch of applicable mathematics, which enables us to consider a large number of unrelated linear and non-linear problems arising in pure and applied sciences in a general and unified framework. In this article, the author shows that the ideas and techniques of variational inequality theory can be extended to study a class of non-linear obstacle problems in physical oceanography. These obstacle boundary value problems originate as a boundary value problem from a model of large scale ocean circulation. Several computational techniques for computing the approximate solutions of the variational inequalities are given. The author also suggests some open problems for further research with additional information and references.

I am indebted to many of my colleagues and students who directly or indirectly helped me in the preparation of this book. I am grateful to Mr Lance Sucharov, Publishing Director and Mrs Juliet Booker, Book Production Editor, at Computational Mechanics Publications for their encouragement and help in improving the presentation and contents of this title. Sincere thanks are due to all authors who contributed to this volume. I must express my gratitude to my family for their patience and understanding during the lengthy preparation of this book.

## References

1. Sir Horace Lamb (1945). *Hydrodynamics*, Dover, New York.

2. Whitham, G.B. (1974). *Linear and Nonlinear Waves*, John Wiley and Sons, New York.

3. Wehausen, J.V. & Laitone, E.V. (1960). *Surface Waves*, Handbuch der Physik, Vol. 9, Springer-Verlag, Berlin.

4. Wiegel, R.E. (1964). *Oceanographical Engineering*, Prentice-Hall, Englewood Cliffs, N.J.

5. Airy, G.B. (1845). *Tides and Waves*, Encyclopaedia Metropolitan, Article 192, London.

Matiur Rahman
Series Editor
Halifax, Canada

# Chapter 1

# Low-frequency asymptotics of hydrodynamic forces on fixed and floating structures


## P. McIver

*Department of Mathematical Sciences, Loughborough University of Technology, Loughborough, Leicestershire LE11 3TU, UK*



## Abstract

This article is concerned with the time-harmonic linearised theory of the interaction of water waves and structures. By making the assumption that the frequency of the motion is sufficiently low, or equivalently that the waves are much longer than a typical body dimension, it is possible to derive simple analytic formulae for the wave forces on structures. These formulae aid the understanding of the complex dependence of the forces on the structural geometry and the wave parameters and may also be used as a partial check on numerical calculations. In this article some of the methods for investigating low-frequency asymptotics are reviewed and a comprehensive set of results is presented for both finite depth water and deep water. The main part of the article illustrates how simple applications of the method of matched asymptotic expansions may be used to obtain simple results for single-element structures. The more complex issue of multi-element structures is also addressed briefly.


## Nomenclature

Some of the most extensively used notation is given below together with the section or equation number where it first appears.

| | |
|---|---|
| $a$ | typical structural dimension [§4] |
| $a_{st}$ | added mass coefficient [eqn (17)] |
| $\hat{a}_{st}$ | leading-order added mass coefficient [eqn (65)] |
| $A$ | amplitude of incident wave [eqn (3)] |
| $\mathcal{A}(\psi)$ | far field angular dependence of diffracted field [eqn (9)] |
| $\mathcal{A}_s(\psi)$ | far field angular dependence of waves radiated in mode $s$ [eqn (10)] |
| $\mathcal{A}_s^{\pm}$ | amplitudes of waves radiated in mode $s$ for 2-D motion [eqn (25)] |
| $b_{st}$ | damping coefficient [eqn (17)] |

| | |
|---|---|
| $\hat{b}_{st}$ | leading-order damping coefficient [eqn (66)] |
| $c_g$ | group velocity [eqn (14)] |
| $d$ | depth of submergence [§6] or gap width [§9.1] |
| $k$ | wavenumber [eqn (3)] |
| $K$ | $(= \omega^2/g)$ deep water wavenumber [eqn (5)] |
| $h$ | water depth [§2.1] |
| $l$ | typical spacing between array elements [§11] |
| $n$ | normal coordinate directed out of fluid [§2.1] |
| $n_m$ | component of normal in direction $m$ [§2.1] |
| $r$ | polar radius in horizontal plane [§2.1] |
| $R$ | 2-D, polar radius in vertical plane [§5] |
| | 3-D, spherical polar coordinate [§7] |
| $\mathcal{R}$ | reflection coefficient [eqn (23)] |
| $\mathcal{T}$ | transmission coefficient [eqn (24)] |
| $V$ | 2-D, submerged cross-sectional area of structure [eqn (78)] |
| | 3-D, submerged volume of structure [eqn (33)] |
| $X_s$ | exciting force in mode $s$ [eqn (11)] |
| $\overline{X}_s$ | mean drift force in mode $s$ [eqns (21-22) and (29)] |
| $\gamma$ | Euler's constant [eqn (53)] |
| $\Gamma$ | wetted surface of isolated structure [§2.1] |
| $\Gamma_j$ | wetted surface of structural element $j$ [§2.1] |
| $\theta$ | 2-D, polar angle in vertical plane [§5] |
| | 3-D, spherical polar coordinate [§7] |
| $\mu_s$ | dipole coefficient for flow in direction $s$ [eqn (71)] |
| $\rho$ | water density [eqn (11)] |
| $\hat{\phi}$ | leading-order outer region potential [§5.1] |
| $\hat{\varphi}$ | leading-order inner region potential [§5.1] |
| $\phi_D$ | diffraction potential [eqn (2)] |
| $\phi_I$ | incident potential [eqn (2)] |
| $\phi_s$ | radiation potential for mode $s$ [eqn (2)] |
| $\psi$ | polar angle in horizontal plane [§2.1] |
| $\omega$ | oscillation frequency [eqn (3)] |

# 1 Introduction

This article is concerned with the wave forces on offshore structures due to a simple time-harmonic forcing resulting from either an incident plane wave or the forced oscillations of the structure. All motions are assumed to have sufficiently small amplitude for the linearised theory to be used and only the contributions to the forces from the inviscid potential flow are considered. A brief review of the linearised theory is given in §2 of this article. In practice, viscous forces will be important in the low-frequency or, equivalently, long wavelength regime discussed here.

The simplest approach to calculating wave exciting forces on ships, or offshore structures, is based on the so-called Froude-Krylov hypothesis in which it is assumed that the pressure field is unaffected by the structure and so can be calculated from the incident wave properties. This approach can be justified provided the wavelength is sufficiently large compared to a suitable structural dimension (Newman, 1977, §7.6).

In the relative motion hypothesis (Newman, 1977, §6.18) the long-wave approximation is extended to include diffraction effects. The velocity field induced by the incident wave potential is assumed to be constant across the body and this allows the diffraction potential to be expressed in terms of radiation potentials corresponding to forced oscillations of the structure. An account of some of the results that follow from the relative motion hypothesis is given in the present §3. It is found that, in this low-frequency approximation, the exciting force can be simply related to the added-mass and damping coefficients. The validity of this approximation in a variety of circumstances is discussed in this article. Simon & Hulme (1986) have extended the relative motion hypothesis to higher order in developing asymptotic expansions of hydrodynamic quantities for axisymmetric structures in deep water.

Another way of obtaining asymptotic information in water-wave problems is through the Kramers-Kronig relations first used in ship-motion problems by Kotik & Mangulis (1962), although these relations have been used in other areas of research for many years. The Kramers-Kronig relations link the added-mass and damping coefficients in terms of integrals over all frequencies. However, Kotik & Mangulis (1962) show how they may be used to obtain information about the low-frequency behaviours of the added mass and damping coefficients and their results are given in the present §4. Greenhow (1986) has extended these ideas to obtain the structure of the asymptotic expansions of hydrodynamic coefficients for both low and high frequency. Further results relevant to low frequencies are described by Greenhow in an appendix to McIver & Linton (1991).

The bulk of the present article is concerned with the systematic development of low-frequency approximations to added mass and damping coefficients, exciting forces and drift forces. The results are derived by the method of matched asymptotic expansions which, in the style used here, arose out of work by Lord Rayleigh in acoustics and was introduced into water waves principally through the work of Tuck on flow through small holes; a review is given by Tuck (1975). Here the derivations are given in terms of an informal matching procedure which appeals strongly to intuition, although all of the results have been verified by the author using a formal application of the method. There are many examples in the literature where erroneous results have arisen through failure to use a formal matching principle. A good recent account of the formal approach as applied to wave problems is by Crighton (1992). Although Crighton's examples are drawn from the field of acoustics there are strong similarities with the water-wave problems discussed here.

A brief general discussion of the method applicable to all of the cases considered is given at the beginning of §5. Each of §§5-11 describes results for a different geometrical situation; there is a little repetition of material so as to make each section reasonably self contained. Results for single structures are given in §§5-10. Two-dimensional, surface-piercing and submerged structures in deep water are considered in §§5-6,while the equivalent results for vertically axisymmetric structures are given in §§7-8. For water of finite depth, surface-piercing structures are considered and two-dimensional and vertically axisymmetric results are given in §9 and §10 respectively. In §9.1, there is some discussion of submerged, two-dimensional structures in finite depth water. Finally, in §11, some results on multi-element structures are discussed.

## 2 Background theory

In this section, relevant background to the linearised theory of the interaction of water waves with structures is described. Much of the notation employed here is based on that used in the wave hydrodynamics text by Mei (1983) where the derivations of the results given below may be found.

### 2.1 Preliminaries

The general results described here are for $N$ independent elements that are each able to move in the three rigid-body translational modes of oscillation; the extension to include rotational modes is straightforward and the interested reader is referred to Mei (1983). An oscillator index is defined by

$$s = 3(j-1) + m, \quad j = 1, 2 \ldots N, \quad m = 1, 2, 3, \tag{1}$$

where $j$ is the number of the element and $m$ is the mode of oscillation; there are a total of $3N$ modes. Cartesian coordinates $(x, y, z)$ and standard cylindrical polar coordinates $(r, \psi, z)$ are used having origin in the mean free surface with the $z$ axis directed vertically downwards. Sway, surge and heave motions parallel to the $x$, $y$ and $z$ directions have indices 1, 2 and 3 respectively. The fluid is taken to be of constant depth $h$.

The results are based on the linearised theory of water waves with all motion assumed to be time-harmonic with angular frequency $\omega$. A velocity potential $\Phi$ exists and the time dependence may be extracted, in a standard way, by writing

$$\Phi = \Re\left\{\left[\phi_I + \phi_D + \sum_{s=1}^{3N} U_s\phi_s\right] e^{-i\omega t}\right\}, \tag{2}$$

where $\Re$ indicates that the real part is to be taken. Here, the complex-valued potentials $\phi_I$ and $\phi_D$ describe the incident and diffracted waves, while $\phi_s$ describes the radiated waves due to oscillations in mode $s$. The complex quantities $\{U_s;\ s = 1, 2, \ldots 3N\}$, are the velocity amplitudes for each mode of motion. Initially, results for fully three-dimensional motion are described. Modifications for two-dimensional motion will be given in §2.3.

For an incident wave of amplitude $A$ and wavenumber $k$ propagating in the direction $\psi = 0$, the potential is

$$\phi_I = -\frac{igA}{\omega\cosh kh}\, e^{ikx}\cosh k(z-h). \tag{3}$$

The frequency $\omega$ and the wavenumber $k$ are related through the dispersion relation

$$\omega^2 = gk\tanh kh. \tag{4}$$

Much of the work described here is concerned with deep water, that is with the limit $kh \to \infty$. For future use it is worth noting the limiting behaviour

$$k \to \frac{\omega^2}{g} \equiv K \quad \text{and} \quad \frac{\cosh k(z-h)}{\cosh kh} \to e^{-Kz} \quad \text{as} \quad kh \to \infty. \tag{5}$$

The boundary value problems that determine the potentials $\phi_D$ and $\phi_s$ are described by Mei (1983, §7.3.1). Each of the $\phi_D$ and $\phi_s$ must satisfy Laplace's equation throughout the fluid region. On the surface of element $j$, denoted here by $\Gamma_j$, the boundary conditions

$$\frac{\partial \phi_D}{\partial n} = -\frac{\partial \phi_I}{\partial n} \quad \text{and} \quad \frac{\partial \phi_s}{\partial n} = n_m \quad \text{on} \quad \Gamma_j \tag{6}$$

must be satisfied, where $n$ is a normal coordinate measured into the structure and $n_m$ is the component of the inward normal to the structure in the direction $m$. On the free surface the usual linearised free-surface condition holds for each potential so that

$$\frac{\partial \phi_D}{\partial z} + K\phi_D = 0 \quad \text{and} \quad \frac{\partial \phi_s}{\partial z} + K\phi_s = 0 \quad \text{on} \quad z = 0 \tag{7}$$

while no flow through the sea bed requires

$$\frac{\partial \phi_D}{\partial z} = \frac{\partial \phi_s}{\partial z} = 0 \quad \text{on} \quad z = h. \tag{8}$$

In the far-field, as $kr \to \infty$, the diffraction potential and each radiation potential reduce to outgoing waves of decreasing amplitude having the forms

$$\phi_D \sim -\frac{igA}{\omega \cosh kh} \left(\frac{2}{\pi kr}\right)^{1/2} \mathrm{e}^{ikr - i\pi/4} \mathcal{A}(\psi) \cosh k(z-h) \tag{9}$$

and

$$\phi_s \sim -\frac{ig}{\omega \cosh kh} \left(\frac{2}{\pi kr}\right)^{1/2} \mathrm{e}^{ikr - i\pi/4} \mathcal{A}_s(\psi) \cosh k(z-h), \tag{10}$$

thus defining the far-field angular dependence $\mathcal{A}(\psi)$ for the diffracted wave field and $\mathcal{A}_s(\psi)$ for the radiated waves.

Many of the results described here are for a single structure. The surface of the structure will then be denoted by $\Gamma$ and the index $s$ may be used interchangeably with $m$.

## 2.2 Hydrodynamic forces

The hydrodynamic forces on the structural elements may be calculated by integrating the pressure over their wetted surfaces. These forces are time-dependent quantities but it is convenient to extract the time dependence in the same manner as was done for the velocity potential in eqn (2). Each time-dependent force is written in the form $\Re\left\{F\,\mathrm{e}^{-i\omega t}\right\}$, where $F$ is a complex quantity containing information about the magnitude and phase of the force.

The exciting force on device $j$ in direction $m$ due to the incident waves is

$$X_s = i\omega\rho \iint_{\Gamma_j} (\phi_I + \phi_D)\, n_m \, dS \tag{11}$$

where the oscillator index $s$ is defined by eqn (1) above and $\rho$ is the water density. As explained by Mei (1983, §7.6.3), the exciting force may be related to the far-field behaviour of the corresponding radiation potential through applications of Green's theorem and it is found that

$$X_s = -i\omega\rho \iint_{S_\infty} \left( \phi_I \frac{\partial \phi_s}{\partial n} - \phi_s \frac{\partial \phi_I}{\partial n} \right) dS. \tag{12}$$

Here, $S_\infty$ is an enclosing vertical, cylindrical surface at large distances from the devices and $n$ is an outward normal coordinate to that surface. Substituting from eqns (3) and (10) and integrating using the method of stationary phase gives, as $kr \to \infty$,

$$X_s = -\frac{4\rho g A c_g}{k} \mathcal{A}_s(\pi), \tag{13}$$

where the group velocity

$$c_g = \frac{\omega}{2k}\left(1 + \frac{2kh}{\sinh 2kh}\right). \tag{14}$$

For future use it is worth noting the asymptotic forms

$$c_g \sim \frac{\omega}{2K} \quad \text{as} \quad kh \to \infty \quad \text{and} \quad c_g \sim \frac{\omega}{k} \quad \text{as} \quad kh \to 0. \tag{15}$$

Thus, the exciting force on the fixed device in direction $s$, $X_s$, is directly related to the disturbance radiated towards the oncoming waves by forced oscillations in mode $s$.

The total hydrodynamic force on body $j$ in direction $m$ due to forced body motions is

$$F_s = i\omega\rho \iint_{\Gamma_j} \left[ \sum_{t=1}^{3N} U_t \phi_t \right] n_m \, dS = i\omega \sum_{t=1}^{3N} U_t \left( a_{st} + i\frac{b_{st}}{\omega} \right), \tag{16}$$

say, where

$$a_{st} = \rho \Re \iint_{\Gamma_j} \phi_t n_m \, dS \quad \text{and} \quad b_{st} = \omega\rho \Im \iint_{\Gamma_j} \phi_t n_m \, dS \tag{17}$$

are the added mass and damping coefficients describing the force in mode $s$ (i.e. on element $j$ in direction $m$) due to forced oscillations of unit velocity amplitude in mode $t$. Here $\Im$ indicates that the imaginary part is to be taken. Note that the added mass and damping obey the symmetry relations $a_{ts} = a_{st}$ and $b_{ts} = b_{st}$. The damping coefficient is related to the waves radiated to the far field and, by Green's theorem, may be written (Mei, 1983, §7.6.1)

$$b_{st} = \frac{i\omega\rho}{2} \iint_{S_\infty} \left( \phi_t \frac{\partial \phi_s^*}{\partial n} - \phi_s^* \frac{\partial \phi_t}{\partial n} \right) dS \tag{18}$$

which on substitution from eqn (10) gives

$$b_{st} = \frac{2\rho g c_g}{\pi k} \int_0^{2\pi} \mathcal{A}_s^*(\psi) \mathcal{A}_t(\psi) \, d\psi, \tag{19}$$

where an asterisk denotes complex conjugate. In addition, the damping coefficients may be related to the exciting forces by substituting for $\mathcal{A}_s(\psi)$ from eqn (13) giving

$$b_{st} = \frac{k}{8\pi\rho g|A|^2 c_g} \int_0^{2\pi} X_s^*(\psi) X_t(\psi)\, d\psi. \tag{20}$$

The linear solution may also be used to calculate the mean second-order, or drift, forces on the body. In particular, momentum arguments (Mei, 1983, §7.10) may be used to show that the mean second-order force on a fixed structure is

$$\overline{X_1} = \frac{\rho g A^2 c_g}{\pi\omega} \int_0^{2\pi} (1 - \cos\psi)|\mathcal{A}(\psi)|^2\, d\psi \tag{21}$$

in the direction of wave advance (that is the $x$ direction) and

$$\overline{X_2} = -\frac{\rho g A^2 c_g}{\pi\omega} \int_0^{2\pi} \sin\psi\, |\mathcal{A}(\psi)|^2\, d\psi \tag{22}$$

in the $y$ direction. Here $\mathcal{A}(\psi)$ is the far-field angular dependence of the diffracted wave field defined by eqn (9).

### 2.3 Results in two-dimensions

In this section modifications of the above results are given when the fluid motion is purely two-dimensional in the plane $y = 0$. The diffraction problem may then be interpreted as normal incidence on a long cylinder with generators parallel to the $y$ axis. Only oscillation modes of this cylinder that radiate plane waves parallel to the $x$ axis are considered.

The decomposition of the potential and the form of the incident wave are still given by eqns (2) and (3) respectively. In the far field the form of the diffracted wave is

$$\phi_D \sim -\frac{igA}{\omega\cosh kh}\mathcal{R}\, \mathrm{e}^{-ikx} \cosh k(z-h) \quad \text{as} \quad x \to -\infty \tag{23}$$

and

$$\phi_D \sim -\frac{igA}{\omega\cosh kh}(\mathcal{T} - 1)\, \mathrm{e}^{ikx} \cosh k(z-h) \quad \text{as} \quad x \to \infty \tag{24}$$

where $\mathcal{R}$ and $\mathcal{T}$ are respectively the reflection and transmission coefficients. The far-field forms of the radiated waves are written

$$\phi_s \sim -\frac{ig}{\omega\cosh kh}\mathcal{A}_s^{\pm}\, \mathrm{e}^{\pm ikx} \cosh k(z-h) \quad \text{as} \quad x \to \pm\infty \tag{25}$$

so that $\mathcal{A}_s^{\pm}$ are the amplitudes of the plane waves radiated to either infinity.

All hydrodynamic forces are now interpreted as forces per unit length of the cylinder and the double integrals in the definitions (11) and (17) become single integrals over the cylinder cross section. By using ideas similar to those for the three-dimensional case above, it may be shown that the exciting forces and damping coefficients are related to the far-field radiated waves by

$$X_s = -2\rho g A c_g \mathcal{A}_s^- \tag{26}$$

and

$$b_{st} = \rho g c_g \left( \mathcal{A}_s^- \mathcal{A}_t^{-*} + \mathcal{A}_s^+ \mathcal{A}_t^{+*} \right), \tag{27}$$

respectively. For a structure that is symmetric about $x = 0$, so that the heights of the waves radiated to the left and right are equal, elimination of the radiated wave amplitudes from (26) and (27) gives the result

$$b_{ss} = \frac{1}{2\rho g c_g} \left| \frac{X_s}{A} \right|^2. \tag{28}$$

Finally, the mean drift force is given in terms of the reflection coefficient as

$$\overline{X_1} = \frac{\rho g A^2 k c_g}{\omega} |\mathcal{R}|^2. \tag{29}$$

## 3 The relative motion hypothesis

The so-called 'relative motion hypothesis' and its consequences are described by Newman (1977, §6.18). The aim is to obtain a low-frequency approximation to the exciting forces on a single, surface-piercing structure. The water is assumed to be infinitely deep, that is the limit $kh \to 0$ is taken; by convention the deep-water wavenumber is denoted by $K$. The equivalent results for finite depth are discussed in later sections.

The origin of coordinates is chosen within the waterplane area of the structure and it is assumed that the wavelength of the incident wave $2\pi/K$ is much greater than a typical dimension $a$ of the structure so that $Ka \ll 1$. With this assumption, in the vicinity of the body the incident wave potential (3), with $kh \to \infty$ as in eqn (5), may be expanded as

$$\phi_I = -\frac{igA}{\omega} \left(1 + iKx - Kz + O\left((Ka)^2\right)\right) \tag{30}$$

so that the body boundary condition (6) for the diffraction potential is approximately

$$\begin{aligned} \frac{\partial \phi_D}{\partial n} &\sim \frac{igA}{\omega} \left( iK\frac{\partial x}{\partial n} - K\frac{\partial z}{\partial n} \right) = \frac{igA}{\omega} (iKn_1 - Kn_3) \\ &= -\omega A \left( \frac{\partial \phi_1}{\partial n} + i\frac{\partial \phi_3}{\partial n} \right) \quad \text{on} \quad \Gamma \end{aligned} \tag{31}$$

where the body boundary conditions for the radiation potentials in eqn (6) have been used to replace the components of the normal. The form of (31) suggests the approximation

$$\phi_D \sim -\omega A (\phi_1 + i\phi_3) \tag{32}$$

for the diffraction potential in the vicinity of the structure. The diffracted field appears to be generated by a combination of sway and heave motions.

The exciting forces are given by eqn (11) with $\phi_I$ replaced by eqn (30) and $\phi_D$ by eqn (32). Consider first the contribution from the incident wave potential. By the divergence theorem, for an arbitrary function $\Psi$

$$\iint_\Gamma \Psi n_s \, dS = -\iiint_V \nabla\Psi \cdot \mathbf{e}_s \, dV - \iint_S \Psi n_s \, dS \tag{33}$$

where $\mathbf{e}_s$ is a unit vector in the direction of mode $s$, $V$ is the submerged volume of the structure and $S$ is the waterplane area in $z = 0$. The reader is reminded that the surface of an isolated structure is denoted by $\Gamma$, the mode index $s$ is used instead of $m$ and that the surface normal is taken to be directed into the structure. The approximation (30) for $\phi_I$ in eqn (33) gives

$$\begin{aligned}\iint_\Gamma \phi_I n_s \, dS &\sim -\iiint_V \left(-\frac{igA}{\omega}\right)(iK\mathbf{e}_1 - K\,\mathbf{e}_3)\cdot \mathbf{e}_s \, dV \\ &\quad - \iint_S \left(-\frac{igA}{\omega}\right)(1 + iKx)\, n_s \, dS \\ &= \frac{igA}{\omega}\left\{K\left(i\delta_{s1} - \delta_{s3}\right)V + \delta_{s3}\left(S + iKS_1\right)\right\}\end{aligned} \tag{34}$$

where

$$S_1 = \iint_S x \, dS \tag{35}$$

and $\delta_{st}$ is the Kronecker delta. The contribution to the exciting force integral from the diffraction potential (32) is

$$\begin{aligned}\iint_\Gamma \phi_D n_s \, dS &\sim \iint_\Gamma (-\omega A)(\phi_1 + i\phi_3)\, n_s \, dS \\ &= -\frac{\omega A}{\rho}\left\{a_{s1} + i\frac{b_{s1}}{\omega} + i\left(a_{s3} + i\frac{b_{s3}}{\omega}\right)\right\}.\end{aligned} \tag{36}$$

Long-wave approximations to the exciting forces now follow from eqns (11), (34) and (36). For a structure symmetric about the plane $x = 0$, $S_1$ is zero and there is no coupling between sway and heave so that the off-diagonal terms of the added-mass and damping matrices appearing in eqn (36) are also zero. In this case the sway exciting force is

$$X_1 = -i\omega^2 A\left(\rho V + a_{11} + i\frac{b_{11}}{\omega}\right) \tag{37}$$

and the heave exciting force is

$$X_3 = -A\left[\rho g S - \omega^2\left(\rho V + a_{33} + i\frac{b_{33}}{\omega}\right)\right]. \tag{38}$$

The relative motion hypothesis always gives a zero force in the horizontal direction perpendicular to the direction of wave advance.

The above informal derivation of eqns (37) and (38) does not indicate explicitly the relative sizes of the various terms. Part of the motivation of the work presented in §§5-10 is to examine these approximations more fully.

## 4 Consequences of the Kramers-Kronig relations

The Kramers-Kronig relations provide a link between the added mass and damping coefficients. Define a complex force coefficient

$$q(\omega) = a_{st}(\omega) - a_{st}(\infty) + i\frac{b_{st}(\omega)}{\omega} \tag{39}$$

where $a_{st}$ and $b_{st}$ are any pair of added mass and damping coefficients. Causality implies that, when regarded as a function of complex frequency $\Omega$, $q(\Omega)$ is analytic in the upper half plane and

$$q^*(-\Omega) = q(\Omega), \tag{40}$$

see Wehausen (1971). Cauchy's theorem applied to $q(\Omega)$, with an integration contour along the real axis and a semicircle of large radius in the upper half plane, and use of eqn (40) yields, after some rearrangement,

$$q(\omega) = \frac{2}{\pi i} \fint_0^\infty \frac{\omega \Re\{q(\Omega)\} + i\Omega \Im\{q(\Omega)\}}{\Omega^2 - \omega^2}\, d\Omega. \tag{41}$$

Now taking real and imaginary parts gives the Kramers-Kronig relations

$$a_{st}(\omega) - a_{st}(\infty) = \frac{2}{\pi} \fint_0^\infty \frac{b_{st}(\Omega)}{\Omega^2 - \omega^2}\, d\Omega \tag{42}$$

and

$$b_{st}(\omega) = -\frac{2\omega^2}{\pi} \fint_0^\infty \frac{a_{st}(\Omega) - a_{st}(\infty)}{\Omega^2 - \omega^2}\, d\Omega. \tag{43}$$

The derivation of eqns (42) and (43) assumes that $q(\omega)$ decays sufficiently quickly as $\omega \to \infty$ and that any singularities on the real axis are sufficiently weak not to contribute to the contour integration. McIver & Linton (1991) investigated the two-dimensional heaving of a structure in finite depth fluid and in this case the complex force coefficient has a simple pole at the origin. The necessary modification to the Kramers-Kronig relations is given by M. Greenhow in an appendix to that paper.

Kotik & Mangulis (1962) discuss a number of asymptotic consequences of the Kramers-Kronig relations two of which are relevant to low frequencies. When discussing these consequences it is more convenient to deal with non-dimensional quantities. To this end, define non-dimensional added mass and damping by

$$\mu(Ka) = \frac{a_{st}(\omega)}{M} \quad \text{and} \quad \lambda(Ka) = \frac{b_{st}(\omega)}{\omega M} \tag{44}$$

where $M$ is a reference mass, $a$ is a typical structural dimension and $K = \omega^2/g$. Kotik & Mangulis (1962) show that, provided $\lambda(0) \neq 0$, then to leading order

$$\mu(Ka) \sim -\frac{\lambda(0)}{\pi} \ln Ka. \tag{45}$$

On the other hand Kotik & Mangulis show that if

$$\lambda(Ka) = \lambda_1 Ka + o(Ka) \quad \text{as} \quad Ka \to 0 \tag{46}$$

then

$$\mu(Ka) \sim \mu(0) - \frac{\lambda_1}{\pi} Ka \ln Ka. \tag{47}$$

## 5 Two-dimensional, surface-piercing structures in deep water

In this and the following three sections informal applications of the method of matched asymptotic expansions are used to derive the hydrodynamic forces on single structures in deep water. For simplicity all the structures considered are symmetric about the plane $x = 0$ although the results may readily be generalised.

The situation to be investigated is when the wavelength $2\pi/K$ is greater than a typical structural dimension $a$ so that

$$Ka << 1. \tag{48}$$

The occurrence of two different length scales is the classic indication of a singular perturbation problem and a single expansion in terms of the small parameter $Ka$ cannot be expected to be valid throughout the fluid domain. To obtain perturbation solutions the flow domain is thought of in terms of two regions. The inner region surrounds the structure to radial distances $r$ much less than the wavelength so that $r \ll K^{-1}$. In the inner region wave effects are not discernible. The outer region is at distances $r$ much greater than the body dimension $a$. Expansions in $Ka$ are constructed in each of the two regions and matched asymptotically in an overlap region which is assumed to exist. The matching requires that the inner expansion of the outer solution coincides with the outer expansion of the inner solution. An intuitive approach will be followed here in preference to giving the formal details of the matching procedures. In the main results, the small parameters are displayed explicitly so that the orders of magnitude of different quantities may be compared directly.

In this section, a long surface-piercing cylinder with horizontal generators parallel to the $y$ axis is considered. The origin of coordinates is chosen at the centre of the intersection of the structure with the free surface so that the waterline occupies $|x| \leq a$. The fluid motion is taken to be two-dimensional in the $x$, $z$-plane. Polar coordinates $(R, \theta)$ are defined by

$$x = R\sin\theta, \quad z = R\cos\theta, \tag{49}$$

so that $\theta$ is measured from the downward vertical.

### 5.1 Heave radiation problem

First of all vertical, or heave, oscillations of the cylinder are considered. In the far field waves radiate symmetrically in the horizontal directions. At low frequencies the primary outer region flow appears to be due to a wave source at the origin and so the leading-order outer solution is written

$$\hat{\phi}_3 = \alpha_3 g_0(x, z). \tag{50}$$

(Throughout this article a 'hat' is used to denote the leading-order low-frequency approximation to a quantity.) Here $\alpha_3$ is a constant to be found and the wave source (Thorne, 1953)

$$g_0 = -\int_0^\infty \frac{\cos\kappa x\, \mathrm{e}^{-\kappa z}}{\kappa - K}\, d\kappa \qquad (51),$$

where the integration path runs beneath the pole in the integrand so as to satisfy the radiation condition. As $K|x| \to \infty$

$$g_0 \sim -\pi i\, \mathrm{e}^{iK|x|}\, \mathrm{e}^{-Kz} \tag{52}$$

giving outgoing waves and as $KR \to 0$

$$g_0 = \ln KR + \gamma - \pi i + O(KR \ln KR) \tag{53}$$

indicating the source-like character. Here $\gamma$ is Euler's constant. Combining eqns (50) and (53), the inner expansion of the outer solution is

$$\hat{\phi}_3 \sim \alpha_3 (\ln KR + \gamma - \pi i) \quad \text{as} \quad KR \to 0 \tag{54}$$

(logarithmic terms and constants must be taken together, see Crighton, 1992).

The leading terms in the inner potential must satisfy the body boundary condition (6) and the low-frequency approximation to the free-surface boundary condition obtained by setting $K = 0$ in (7). Let $\chi_3$ be a harmonic function that satisfies

$$\frac{\partial \chi_3}{\partial n} = n_3 \quad \text{on} \quad \Gamma \quad \text{and} \quad \frac{\partial \chi_3}{\partial z} = 0 \quad \text{on} \quad z = 0. \tag{55}$$

To match with (53), $\chi_3$ is chosen to have the far field behaviour

$$\chi_3 - \alpha_3 \ln(R/a) \to 0 \quad \text{as} \quad R/a \to \infty. \tag{56}$$

Any constant term can be added to the inner potential without violating the boundary conditions so the leading order solution in the inner region is written

$$\hat{\varphi}_3 = C + \chi_3(x, z). \tag{57}$$

The asymptotic form for $\chi_3$ in eqn (56) gives the outer expansion of the inner solution as

$$\hat{\varphi}_3 \sim C + \alpha_3 \ln(R/a) \quad \text{as} \quad R/a \to \infty. \tag{58}$$

Matching this with the inner expansion of the outer solution in eqn (54) gives

$$C = \alpha_3 \{\ln Ka + \gamma - \pi i\} \tag{59}$$

and so the leading-order inner solution

$$\hat{\varphi}_3 = \alpha_3 \{\ln Ka + \gamma - \pi i\} + \chi_3(x, z). \tag{60}$$

The constant $\alpha_3$ is determined by equating the flow out of the inner region due to $\chi_3$, with that generated by the body boundary condition. Let $S_\infty$ be a semicircular contour enclosing the body then the flow out of the inner region

$$\int_{S_\infty} \frac{\partial \hat{\varphi}_3}{\partial R} dS = \lim_{R/a \to \infty} \int_{-\pi/2}^{\pi/2} \frac{\partial \chi_3}{\partial R} R \, d\theta = \alpha_3 \pi \tag{61}$$

where the asymptotic behaviour (59) has been used. The flow into the inner region across the body surface is

$$\int_{\Gamma} \left( -\frac{\partial \hat{\varphi}_3}{\partial n} \right) dS = -\int_{\Gamma} n_3 \, dS = 2a \tag{62}$$

where the last step follows from the two-dimensional equivalent of eqn (33) with $\Psi \equiv 1$. Equating these last two expressions gives

$$\alpha_3 = 2a/\pi. \tag{63}$$

The heave added mass and damping follow from (17) and the inner solution (60) so that to leading order

$$a_{33} + i\frac{b_{33}}{\omega} = \rho \int_{\Gamma} \phi_3 n_3 \, dS \sim \rho \int_{\Gamma} \left\{ \frac{2a}{\pi} (\ln Ka + \gamma - \pi i) + \chi_3 \right\} n_3 \, dS. \tag{64}$$

That is, as $Ka \to 0$,

$$a_{33} \sim \hat{a}_{33} = \rho \left\{ -\frac{4a^2}{\pi} (\ln Ka + \gamma) + \int_{\Gamma} \chi_3 n_3 \, dS \right\} \tag{65}$$

and

$$b_{33} \sim \hat{b}_{33} = 4\rho a^2 \omega \tag{66}$$

where the result in eqn (62) has been used to evaluate an integral.

The damping coefficient may also be found from eqn (27) using the far-field form of the radiation potential

$$\hat{\phi}_3 \sim -2ai \, e^{iK|x|} \, e^{-Kz} \quad \text{as} \quad K|x| \to \infty. \tag{67}$$

Comparison of (67) with eqn (25), in the limit $kh \to \infty$, gives

$$\mathcal{A}_3^+ = \mathcal{A}_3^- = \frac{2}{\omega} Ka \tag{68}$$

which when inserted in (27) recovers (66). The form of the damping in eqn (66) also satisfies the criterion needed for application of the result (45) derived from the Kramers-Kronig relations. Thus,the dominant term in eqn (65) could have been obtained more economically by calculating the damping by the far-field method and then applying eqn (45).

### 5.2 Sway radiation problem

The cylinder is now forced to oscillate in sway and the leading order inner solution is written

$$\hat{\varphi}_1 = \chi_1(x, z) \tag{69}$$

where $\chi_1$ must satisfy the body boundary condition (6) and the low frequency approximation to the free-surface condition (7), that is

$$\frac{\partial \chi_1}{\partial n} = n_1 \quad \text{on} \quad \Gamma \quad \text{and} \quad \frac{\partial \chi_1}{\partial z} = 0 \quad \text{on} \quad z = 0. \tag{70}$$

The boundary conditions (70) indicate that the potential $\chi_1$ is just the disturbance that would be generated by the structure, and its reflection in $z = 0$, when moving uniformly in the $x$ direction. In the far field $\chi_1$ is expected to satisfy (Newman, 1977, §4.14)

$$\chi_1 + \mu_1 a^2 x/R^2 \to 0 \quad \text{as} \quad R/a \to \infty, \tag{71}$$

where $\mu_1$ is the dipole coefficient for the double-body flow. A non-zero constant term might be added to the inner solution without violating the boundary conditions in eqns (70) but this cannot be matched with the outer solution and so does not appear in eqn (69).

In order to match with the form (71), the leading-order approximation to the outer flow must be due to a wave dipole at the origin and so is written

$$\hat{\phi}_1 = \alpha_1 g_1(x, z), \tag{72}$$

where $\alpha_1$ is a constant to be found and the wave dipole (Thorne, 1953)

$$g_1 = \frac{1}{K}\frac{\partial g_0}{\partial x} = \frac{1}{K}\int_0^\infty \frac{\kappa \sin \kappa x \, \mathrm{e}^{-\kappa z}}{\kappa - K}\, d\kappa. \tag{73}$$

As $K|x| \to \infty$

$$g_1 \sim \pi \operatorname{sgn} x \, \mathrm{e}^{iK|x|}\, \mathrm{e}^{-Kz}, \tag{74}$$

so that waves propagate away antisymmetrically, and when $KR \to 0$

$$g_1 = \frac{x}{KR^2} + O(KR \ln KR). \tag{75}$$

Matching the inner and outer solutions (69) and (72), with their asymptotic forms given by (71) and (75) respectively, requires

$$\alpha_1 = -\mu_1 K a^2. \tag{76}$$

The leading-order inner solution is real, so from eqn (17) can only yield an approximation to the added mass. It follows from eqns (17) and (69) that

$$a_{11} \to \hat{a}_{11} = \rho \int_\Gamma \chi_1 n_1 \, dS \quad \text{as} \quad Ka \to 0. \tag{77}$$

The zero-frequency added mass $\hat{a}_{11}$ is directly proportional to the added mass of the double body moving uniformly through an infinite fluid and this added mass may be related to the dipole coefficient (Newman, 1977, p. 144). After adjustment by a factor of two to allow for the double body, it is found that

$$\hat{a}_{11} = \rho(\pi \mu_1 a^2 - V), \tag{78}$$

where $V$ is here the submerged cross-sectional area of the structure (that is, the submerged volume per unit length). This result may also be derived by an application of Green's theorem to $\chi_1$ and $\Psi \equiv x$ over the double body and an enclosing circle at large distances.

To obtain the leading-order approximation to the damping without extending the solutions for the radiation potential to higher order, the far-field relation (27) may be used in a similar way to the calculation at the end of §5.1. It follows from eqns (72), (74) and (25) that the amplitudes of the waves radiated to the far field are

$$A_1^+ = -A_1^- = -\frac{\pi i \mu_1}{\omega}(Ka)^2 \tag{79}$$

so that

$$b_{11} \sim \hat{b}_{11} = \rho \pi^2 a^2 \mu_1^2 \omega (Ka)^2. \tag{80}$$

**5.3 Diffraction problem**

Within distances $O(a)$ of the structure, the deep water incident wave

$$\phi_I = -\frac{igA}{\omega} e^{iKx} e^{-Kz} \sim -\frac{igA}{\omega}(1 + iK(x + iz)) \tag{81}$$

so that the body boundary condition (6) for the diffracted field may be approximated as

$$\frac{\partial \phi_D}{\partial n} = -\frac{\partial \phi_I}{\partial n} \sim -\omega A \frac{\partial}{\partial n}(x + iz) = -\omega A(n_1 + in_3). \tag{82}$$

As has been already shown in §3 on the relative motion hypothesis, the low-frequency diffraction potential is related to the radiation potentials in heave and sway. From the results of §§5.1-2, the boundary and matching conditions are satisfied to leading order by an inner region potential combining the heave and sway radiation potentials, so that,

$$\hat{\varphi}_D = -\omega A \left\{ \chi_1(x, z) + i \left[ \frac{2a}{\pi}(\ln Ka + \gamma - \pi i) + \chi_3(x, z) \right] \right\} \tag{83}$$

and an outer region potential

$$\hat{\phi}_D = -\omega A i \frac{2a}{\pi} g_0(x, z). \tag{84}$$

The leading order diffraction potential contains only the source term as the leading terms in the expansion of (83) as $R/a \to \infty$ come from the heave potential. The sway exciting force arises from those terms in (83) that are antisymmetric in $x$ so that, from eqn (11),

$$X_1 = i\omega\rho \int_\Gamma (\phi_I + \phi_D) n_1 \, dS \sim i\omega^2 \rho A \int_\Gamma (x - \chi_1) n_1 \, dS = -\frac{igA}{a} Ka\,(\rho V + \hat{a}_{11}) \tag{85}$$

while the heave exciting force arises from the symmetric terms and is

$$\begin{aligned} X_3 &= i\omega\rho \int_\Gamma (\phi_I + \phi_D) n_1 \, dS \\ &\sim \omega\rho A \int_\Gamma \left\{ \frac{g}{\omega}(1 - Kz) + \omega \left[ \frac{2a}{\pi}(\ln Ka + \gamma - \pi i) + \chi_3 \right] \right\} n_3 \, dS \\ &= -2\rho g a A \left\{ 1 - \frac{Ka}{2\rho a^2} \left( \rho V + \hat{a}_{33} + i \frac{\hat{b}_{33}}{\omega} \right) \right\}. \end{aligned} \tag{86}$$

Integrations in the derivations of (85-6) are carried out using the result in eqn (33). Comparison of (85) for the sway force with the result (37) from the relative motion hypothesis shows that the consistent leading-order approximation to the exciting force should not contain the damping term. Without further detailed analysis it is unclear whether the damping term constitutes the next term in the low-frequency expansion of the exciting force. On the other hand, the expression (86) for the heave force agrees exactly with eqn (38) derived in §3.

In the heave case, the Haskind relation (28) may be used to obtain further terms in the low-frequency expansion of the damping coefficient as suggested by Greenhow & Ahn (1988). Substituting (86) into (28) and retaining terms up to order $Ka$ (the highest order terms which may be retained consistently) gives

$$b_{33} \sim 4\rho a^2 \omega \left\{ 1 - \frac{Ka}{\rho a^2} \left( \rho V + \hat{a}_{33} \right) \right\}. \tag{87}$$

Approximations to the reflection and transmission coefficients follow from eqns (23-4) and (84), with the asymptotic behaviour of the wave source given by (52). It is found that

$$\mathcal{R} \sim -2iKa \quad \text{and} \quad \mathcal{T} \sim 1 - 2iKa. \tag{88}$$

From (29), the horizontal drift force is therefore

$$\overline{X_1} \sim 2\rho g A^2 (Ka)^2 \tag{89}$$

which is independent of the submerged shape of the structure.

### 5.4 The half-immersed circular cylinder

To further illustrate the results obtained in the present section application is made to a half-immersed circular cylinder of radius $a$. In this case, the boundary conditions (55) to be satisfied by the harmonic heave potential $\chi_3$ become

$$\frac{\partial \chi_3}{\partial R} = \cos\theta \quad \text{on} \quad R = a \quad \text{and} \quad \frac{\partial \chi_3}{\partial \theta} = 0 \quad \text{on} \quad |\theta| = \pi/2,\ |x| > a. \tag{90}$$

It is easily verified that a particular solution with the far-field behaviour (56) is

$$\chi_3 = \frac{2a}{\pi} \left( \ln \frac{R}{a} + \sum_{n=1}^{\infty} \frac{(-1)^n}{n(4n^2-1)} \frac{\cos 2n\theta}{(R/a)^{2n}} \right) \tag{91}$$

and that

$$\begin{aligned} \int_\Gamma \chi_3 n_3 \, dS &= \int_{-\pi/2}^{\pi/2} \chi_3 \big|_{R=a} (-\cos\theta) a \, d\theta \\ &= \frac{4a^2}{\pi} \sum_{n=1}^{\infty} \frac{1}{n(4n^2-1)^2} = \frac{4a^2}{\pi} \left( \tfrac{3}{2} - 2\ln 2 \right) \end{aligned} \tag{92}$$

where the series summation is given by Gradshteyn & Ryzhik (1980, p. 8). From eqn (65), the asymptotic behaviour of the heave added mass is therefore

$$\hat{a}_{33} = \frac{4\rho a^2}{\pi} \left( -\ln Ka + \tfrac{3}{2} - 2\ln 2 - \gamma \right) \tag{93}$$

in agreement with Ursell (1964, eqn (A1.18)). Approximations to the heave exciting force and damping now follow from eqns (86) and (87) respectively.

The boundary conditions to be satisfied by the sway potential $\chi_1$ are

$$\frac{\partial \chi_1}{\partial R} = \sin\theta \quad \text{on} \quad R = a \quad \text{and} \quad \frac{\partial \chi_1}{\partial \theta} = 0 \quad \text{on} \quad |\theta| = \pi/2,\ |x| > a \tag{94}$$

and the solution with the far-field behaviour (71) is

$$\chi_1 = -a^2 \frac{\sin\theta}{R} \tag{95}$$

so that the dipole coefficient $\mu_1 = 1$. It follows immediately from eqns (78), (80) and (85) that the limiting value of the sway added mass is

$$\hat{a}_{11} = \tfrac{1}{2}\rho\pi a^2 \tag{96}$$

and the asymptotic behaviours of the sway damping and exciting force are respectively

$$b_{11} \sim \rho\pi^2 a^2 \omega\,(Ka)^2 \tag{97}$$

and

$$X_1 \sim -i\rho g\pi a A\,Ka. \tag{98}$$

## 6 Two-dimensional, submerged structures in deep water

General comments made at the beginning of §5 also apply here. However, in this section, a long submerged cylinder with horizontal generators parallel to the $y$ axis is considered. With a submerged structure there is an additional length scale in the problem, namely the depth of submergence $d$. Interest is still in the case of long waves relative to the body dimension, so that $Ka \ll 1$, but two cases must be distinguished for the magnitude of the submergence $d$ relative to the structural dimension $a$. The more difficult case is $d/a = O(1)$, so that the structure is quite close to the surface and for the solution by matched asymptotic expansions part of the free surface is within the inner region. The scattering problem for a submerged circular cylinder has been solved by Davis & Leppington (1985) using matched asymptotic expansions with the assumption $d/a = O(1)$.

Here attention will be restricted to the case where the body dimension $a$ is much smaller than all other length scales of the problem. In particular, $d/a \gg 1$ so that the cylinder is deeply submerged. No restriction is placed on the wavelength relative to the submergence so that $Kd$ is arbitrary. This means that the solution presented here is strictly a small body solution in which the structural dimension $a$ is allowed to become arbitrarily small compared to other lengths. In the related context of electromagnetic-wave scattering, this problem has been investigated by Davis & Leppington (1977). Also, an outline of the solution for wave diffraction by a deeply submerged circular cylinder, found by matched asymptotic expansions, is given by Leppington (1992).

The structure is assumed to be symmetrically placed about $x = 0$ and the point $(x, z) = (0, d)$ is chosen as a reference point within the structure. Throughout this section, polar coordinates $(R, \theta)$ relative to the structure defined by

$$x = R\sin\theta, \quad z = d + R\cos\theta \tag{99}$$

will be used.

### 6.1 Heave radiation problem

The free-surface no longer appears in the inner region, so the inner solution must satisfy only the body boundary condition and be able to match with the inner expansion of the outer solution. Thus, the leading-order inner solution for the heave problem is written

$$\hat{\varphi}_3 = \chi_3(x, z), \tag{100}$$

where

$$\frac{\partial \chi_3}{\partial n} = n_3 \quad \text{on} \quad \Gamma \tag{101}$$

so that $\chi_3$ is the disturbance generated by uniform motion of the structure in the $z$ direction. In the far-field of the inner region

$$\chi_3 + \mu_3 a^2 (z - d)/R^2 \rightarrow 0 \quad \text{as} \quad R/a \rightarrow \infty \tag{102}$$

where $\mu_3$ is the dipole coefficient for the vertical motion. As in §5.2, when the solution is dipole-like no additional constants are included in the inner solution. To match with (100) the leading-order outer solution is represented with a vertical wave dipole, singular at $(x, z) = (0, d)$, so that the leading-order outer potential is written

$$\hat{\phi}_3 = \alpha_3 g_{13}(x, z), \tag{103}$$

where the vertical wave dipole (Thorne, 1953)

$$g_{13} = \frac{\cos\theta}{KR} - \frac{1}{K}\int_0^\infty \frac{\kappa + K}{\kappa - K} \cos \kappa x \; \mathrm{e}^{-\kappa(z+d)} \; d\kappa. \tag{104}$$

As $K|x| \rightarrow \infty$

$$g_{13} \sim -2\pi i \, \mathrm{e}^{iK|x|} \, \mathrm{e}^{-K(z+d)} \tag{105}$$

and when $KR \rightarrow 0$

$$g_{13} = \frac{\cos\theta}{KR} + O(1). \tag{106}$$

Matching the inner and outer potentials (100) and (103), using the asymptotic forms given by (102) and (106) respectively, gives

$$\alpha_3 = -\mu_3 K a^2. \tag{107}$$

The leading-order inner solution is real valued so eqn (17) yields only the limit of the added mass, namely

$$a_{33} \rightarrow \hat{a}_{33} = \rho \int_\Gamma \chi_3 n_3 \, dS \quad \text{as} \quad a/d \rightarrow 0. \tag{108}$$

The limiting added mass $\hat{a}_{33}$ is that of the structure moving uniformly through an infinite bath of fluid and is related to the dipole coefficient $\mu_3$ (Newman, 1977, p. 144) by

$$\hat{a}_{33} = \rho(2\pi\mu_3 a^2 - V) \tag{109}$$

where $V$ is the cross-sectional area (that is, the submerged volume per unit length).

The leading-order approximation to the damping follows from knowledge of the far-field wave amplitudes. From eqns (103), (105) and (25), these amplitudes are

$$\mathcal{A}_3^+ = \mathcal{A}_3^- = -\frac{2\pi\mu_3\,\mathrm{e}^{-Kd}}{\omega}(Ka)^2 \tag{110}$$

so that by eqn (27) the damping coefficient

$$b_{33} \sim \hat{b}_{33} = 4\rho\pi^2 a^2 \mu_3^2\,\omega\,(Ka)^2\,\mathrm{e}^{-2Kd} \quad \text{as} \quad a/d \to 0. \tag{111}$$

### 6.2 Sway radiation problem

For a submerged body and with the assumptions stated at the beginning of §6, the sway radiation problem is very similar to the heave radiation problem but with the dipole oriented horizontally rather that vertically. The outer flow is described to leading order by

$$\hat{\phi}_1 = \alpha_1 g_{11}(x, z) \tag{112}$$

where the horizontal dipole (Thorne, 1953)

$$g_{11} = \frac{\sin\theta}{KR} + \frac{1}{K}\int_0^\infty \frac{\kappa + K}{\kappa - K}\sin\kappa x\;\mathrm{e}^{-\kappa(z+d)}\;d\kappa. \tag{113}$$

As $K|x| \to \infty$

$$g_{11} \sim 2\pi\,\mathrm{sgn}\,x\;\mathrm{e}^{iK|x|}\,\mathrm{e}^{-K(z+d)} \tag{114}$$

and when $KR \to 0$

$$g_{11} = \frac{\sin\theta}{KR} + O(KR).$$

The leading-order inner solution is the flow resulting from uniform horizontal motion so

$$\hat{\varphi}_1 = \chi_1(x, z) \tag{115}$$

where

$$\chi_1 + \mu_1 a^2 x/R^2 \to 0 \quad \text{as} \quad R/a \to \infty. \tag{116}$$

Matching the asymptotic forms of the leading order inner and outer solutions gives

$$\alpha_1 = -\mu_1 K a^2 \tag{117}$$

and hence as $a/d \to 0$ the limiting added mass is

$$a_{11} \to \hat{a}_{11} = \rho(2\pi\mu_1 a^2 - V) \tag{118}$$

and the damping coefficient

$$b_{11} \sim \hat{b}_{11} = 4\rho\pi^2 a^2 \mu_1^2\,\omega\,(Ka)^2\,\mathrm{e}^{-2Kd}. \tag{119}$$

### 6.3 Diffraction problem

Within distances $O(a)$ of the structure, the incident wave

$$\phi_I = -\frac{igA}{\omega}\,\mathrm{e}^{iKx}\,\mathrm{e}^{-Kz} \sim -\frac{igA}{\omega}\,\mathrm{e}^{-Kd}(1 + iK(x + i(z - d))) \qquad (120)$$

so that the body boundary condition (6) may be approximated as

$$\frac{\partial \phi_D}{\partial n} = -\frac{\partial \phi_I}{\partial n} \sim -\omega A\,\mathrm{e}^{-Kd}(n_1 + in_3). \qquad (121)$$

As in the surface-piercing case §5.3, the inner solution to leading order for the diffraction potential combines the heave and sway radiation potentials so that

$$\hat{\varphi}_D = -\omega A\,\mathrm{e}^{-Kd}\left\{\chi_1(x, z) + i\chi_3(x, z)\right\}. \qquad (122)$$

Similarly, the leading-order outer potential combines the outer radiation potentials giving

$$\hat{\phi}_D = \omega A K a^2\,\mathrm{e}^{-Kd}\left\{\mu_1 g_{11} + i\mu_3 g_{13}\right\}. \qquad (123)$$

The exciting forces follow from eqn (11) and

$$X_1 \sim i\omega^2 \rho A\,\mathrm{e}^{-Kd}\int_\Gamma (x - \chi_1) n_1\,dS = -\frac{igA}{a} Ka\,\mathrm{e}^{-Kd}(\rho V + \hat{a}_{11}) \qquad (124)$$

while

$$X_3 \sim i\omega^2 \rho A\,\mathrm{e}^{-Kd}\int_\Gamma i(z - d - \chi_3) n_3\,dS = \frac{gA}{a} Ka\,\mathrm{e}^{-Kd}(\rho V + \hat{a}_{33}). \qquad (125)$$

Eqns (124) and (125) are the equivalent of the relative motion hypothesis for the case of a deeply submerged body, note the absence of a term involving the damping in these leading-order approximations.

Reflection and transmission coefficients $\mathcal{R}$ and $\mathcal{T}$ follow from the far-field form of the leading-order outer solution in eqn (123), using eqns (105) and (114), so that

$$\mathcal{R} \sim -2\pi i (Ka)^2\,\mathrm{e}^{-2Kd}(\mu_1 - \mu_3) \qquad (126)$$

and

$$\mathcal{T} = 1 + 2\pi i (Ka)^2\,\mathrm{e}^{-2Kd}(\mu_1 + \mu_3). \qquad (127)$$

An approximation to the mean horizontal drift force follows from eqn (29).

### 6.4 The submerged circular cylinder

For a circular cylinder the dipole coefficients corresponding to uniform motion in the vertical and horizontal directions are each equal to unity. This gives immediately the leading-order approximations

$$\hat{a}_{11} = \hat{a}_{33} = \rho\pi a^2, \qquad (128)$$

$$\hat{b}_{11} = \hat{b}_{33} = 4\rho\pi^2 a^2 \omega\,(Ka)^2\,\mathrm{e}^{-2Kd}, \qquad (129)$$

and

$$X_1 = -iX_3 = -2i\rho g\pi a A\,Ka\,\mathrm{e}^{-Kd}. \qquad (130)$$

By eqn (126) the leading-order reflection coefficient is zero which is consistent with the known result that there is no reflection from a submerged circular cylinder in deep water for any frequency of incident wave (Ursell, 1950).

## 7 Axisymmetric, surface-piercing structures in deep water

General comments made at the beginning of §5 apply here also. In this section, axisymmetric surface-piercing structures in deep water are considered. The origin of coordinates is chosen at the intersection of the structures axis of symmetry with the free surface. In addition to the Cartesian coordinates $(x, y, z)$ and cylindrical polar coordinates $(r, \psi, z)$ introduced in §2.1, spherical polar coordinates $(R, \theta, \psi)$ defined through

$$x = R\sin\theta\cos\psi, \quad y = R\sin\theta\sin\psi, \quad z = R\cos\theta \tag{131}$$

will be employed with $0 \le \theta \le \pi/2$ and $0 \le \psi \le 2\pi$. The waterplane area $S$ occupies the disc $r \le a$ in $z = 0$.

### 7.1 Heave radiation problem

The heave radiation problem for an axisymmetric structure is similar to that described in §5.1 for a two-dimensional flow in that, for the far field, the primary effect at low frequencies is described by a wave source at the origin that radiates waves uniformly in all horizontal directions. Thus, the leading-order term in the outer solution is

$$\hat{\phi}_3 = \alpha_3 Ka\, G_0(x, y, z) \tag{132}$$

where the wave source (Hulme, 1982)

$$G_0 = \frac{1}{K} \int_0^\infty \frac{\kappa}{\kappa - K} J_0(\kappa r)\, \mathrm{e}^{-\kappa z}\, d\kappa \tag{133}$$

where $J_m$ denotes the Bessel function of the first kind and order $m$. The constant $\alpha_3$ in eqn (132) is to be determined from matching with the inner solution, the factor $Ka$ has been introduced for convenience and could be incorporated into $\alpha_3$. Hulme (1982) gives the asymptotic forms

$$G_0 \sim \left(\frac{2\pi}{Kr}\right)^{1/2} \exp(-Kz + i(Kr + \pi/4)) \quad \text{as} \quad Kr \to \infty \tag{134}$$

and

$$G_0 = \frac{1}{KR} + O(1) \quad \text{as} \quad KR \to 0. \tag{135}$$

The leading-order inner solution is written

$$\hat{\varphi}_3 = \chi_3 \tag{136}$$

where $\chi_3$ satisfies the body boundary condition (6) and the low-frequency approximation to the free surface condition (7), that is

$$\frac{\partial \chi_3}{\partial n} = n_3 \quad \text{on} \quad \Gamma \quad \text{and} \quad \frac{\partial \chi_3}{\partial \theta} = 0 \quad \text{on} \quad \theta = \pi/2,\ r > a, \tag{137}$$

where

$$\chi_3 - \alpha_3 a/R \to 0, \quad R/a \to \infty. \tag{138}$$

No constant term can be included in the inner solution in order to match with the inner expansion of the outer solution given by eqns (132) and (135).

The constant $\alpha_3$ is found by equating the flow out of the inner region due to $\chi_3$ with that generated by the body boundary condition. Let $S_\infty$ be a hemispherical surface within the fluid at the limits of the inner region. The flow out of the inner region

$$\iint_{S_\infty} \frac{\partial \hat{\phi}_3}{\partial R}\, dS = \lim_{R/a\to\infty} \int_0^{2\pi} \left\{ \int_0^{\pi/2} \frac{\partial \chi_3}{\partial R} R^2 \sin\theta \, d\theta \right\} d\psi = -2\pi\alpha_3 a \tag{139}$$

where the asymptotic behaviour (138) has been used. The instantaneous flow into the inner region across the mean body surface

$$\iint_\Gamma \left(-\frac{\partial \chi_3}{\partial n}\right) dS = -\iint_\Gamma n_3 \, dS = \pi a^2 \tag{140}$$

where the last step follows from eqn (33) with $\Psi \equiv 1$. Equating the last two results gives

$$\alpha_3 = -\tfrac{1}{2}a. \tag{141}$$

As $\chi_3$ is real-valued, the leading-order inner solution (136) in conjunction with eqn (17) yields only the limiting added mass

$$a_{33} \to \hat{a}_{33} = \rho \iint_\Gamma \chi_3 n_3 \, dS \quad \text{as} \quad Ka \to 0. \tag{142}$$

The leading-order damping may be found from the far-field behaviour of the wave field which follows from eqns (132) and (134) and is

$$\hat{\phi}_3 \sim -\tfrac{1}{2}Ka^2 \left(\frac{2\pi}{Kr}\right)^{1/2} \exp(-Kz + i(Kr + \pi/4)) \quad \text{as} \quad Kr \to \infty \tag{143}$$

so that the far-field dependence defined by eqn (9) is

$$\mathcal{A}_3(\psi) = \frac{\pi}{2\omega}(Ka)^2.$$

The damping coefficient now follows from eqn (19) and is

$$b_{33} \sim \hat{b}_{33} = \tfrac{1}{2}\rho\pi^2 a^3 \omega \, Ka \quad \text{as} \quad Ka \to 0. \tag{144}$$

This may now be used in eqn (47) to obtain the next term in the added mass expansion (see also Hulme, 1982, eqn 3.16).

### 7.2 Sway radiation problem

The structure is now forced to oscillate in sway and the leading-order approximation to the inner region potential

$$\hat{\varphi}_1 = \chi_1 \tag{145}$$

must satisfy the body and free-surface conditions

$$\frac{\partial \chi_1}{\partial n} = n_1 \quad \text{on} \quad \Gamma \quad \text{and} \quad \frac{\partial \chi_1}{\partial \theta} = 0 \quad \text{on} \quad \theta = \pi/2,\ r > a \tag{146}$$

so that $\chi_1$ is the disturbance due to uniform horizontal motion by the structure and its reflection in the free surface. This disturbance potential will satisfy

$$\chi_1 + \mu_1 a^3 x / R^3 \to 0 \quad \text{as} \quad R/a \to \infty \tag{147}$$

where $\mu_1$ is the dipole coefficient for the double-body flow. The asymptotic behaviour in eqn (147) is that of a horizontal dipole and so to effect the matching the leading-order outer region flow must be a horizontal wave dipole so that

$$\hat{\phi}_1 = \alpha_1 G_1(x, y, z) \tag{148}$$

where (Hulme, 1982)

$$G_1 = \frac{\cos\psi}{K^2} \int_0^\infty \frac{\kappa^2}{\kappa - K} J_1(\kappa r)\, \mathrm{e}^{-\kappa z}\, d\kappa. \tag{149}$$

As $Kr \to \infty$

$$G_1 \sim \left( \frac{2\pi}{Kr} \right)^{1/2} \exp(-Kz + i(Kr - \pi/4)) \cos\psi \tag{150}$$

and as $KR \to 0$

$$G_1 = -\frac{\sin\theta \cos\psi}{(KR)^2} + O(1/KR). \tag{151}$$

Matching the inner and outer solutions (145) and (148), using the asymptotic forms (147) and (151) respectively, gives

$$\alpha_1 = \mu_1 K^2 a^3. \tag{152}$$

As in the previous section, the leading-order inner solution (145) yields only the limiting added mass

$$a_{11} \to \hat{a}_{11} = \rho \iint_\Gamma \chi_1 n_1\, dS \quad \text{as} \quad Ka \to 0. \tag{153}$$

The added mass of the double-body flow may be related to the dipole coefficient (Newman 1977, p.143) so that, after adjustment by a factor of two to allow for the double body,

$$\hat{a}_{11} = \rho(2\pi\mu_1 a^3 - V) \tag{154}$$

where $V$ is the submerged volume of the structure. The damping coefficient again follows from the far-field potential

$$\hat{\phi}_1 \sim \mu_1 K^2 a^3 \left(\frac{2\pi}{Kr}\right)^{1/2} \exp(-Kz + i(Kr - \pi/4)) \cos\psi \quad \text{as} \quad Kr \to \infty \tag{155}$$

which, when compared with eqn (10), yields

$$A_1(\psi) = \frac{i\pi\mu_1}{\omega}(Ka)^3 \cos\psi \tag{156}$$

and so from eqn (19)

$$b_{11} \sim \hat{b}_{11} = \rho\pi^2 a^3 \mu_1^2 \omega\,(Ka)^3 \quad \text{as} \quad Ka \to 0. \tag{157}$$

This may now be used in eqn (47) to obtain the next term in the low-frequency expansion of the added mass.

### 7.3 Diffraction problem

Within distances $O(a)$ of the structure, the incident wave

$$\phi_I = -\frac{igA}{\omega} e^{iKx} e^{-Kz} \sim -\frac{igA}{\omega}(1 + iK(x + iz)) \tag{158}$$

so that the body boundary condition (6) for the diffracted field may be approximated as

$$\frac{\partial\phi_D}{\partial n} = -\frac{\partial\phi_I}{\partial n} \sim -\omega A \frac{\partial}{\partial n}(x + iz) = -\omega A(n_1 + in_3). \tag{159}$$

From the results of §§7.1-2, this boundary condition is satisfied by a leading-order inner solution

$$\hat{\varphi}_D = -\omega A(\chi_1 + i\chi_3) \sim i\omega A a^2/2R \quad \text{as} \quad R/a \to \infty, \tag{160}$$

where the asymptotic forms (138) and (147) have been used. This source-like behaviour indicates that the leading-order term in the outer solution will be a wave source so that

$$\hat{\phi}_D = \frac{i\omega^3 A a^2}{2g} G_0(x, y, z), \tag{161}$$

where eqn (132), in conjunction with eqn (160), has been used to determine the multiplicative constant. Approximations to the exciting forces follow from eqn (11) and are

$$X_1 \sim i\omega\rho \iint_\Gamma \omega A(x - \chi_1) n_1\, dS = -\frac{-igA}{a} Ka\,(\rho V + \hat{a}_{11}) \tag{162}$$

and

$$\begin{aligned} X_3 &\sim i\omega\rho \iint_\Gamma \left\{-\frac{igA}{\omega}(1 - Kz) - i\omega A\chi_3\right\} n_3\, dS \\ &= -\rho g\pi a^2 A\left\{1 - \frac{Ka}{\rho\pi a^3}(\rho V + \hat{a}_{33})\right\} \end{aligned} \tag{163}$$

which may be compared with the relative motion hypothesis results in eqns (37) and (38).

In the heave case, the exciting force may be substituted into the Haskind relation (20) to obtain the improved estimate for the damping

$$b_{33} \sim \tfrac{1}{2}\rho\pi^2 a^3 \omega\, Ka \left(1 - \frac{2Ka}{\rho\pi a^3}(\rho V + \hat{a}_{33})\right). \tag{164}$$

The far-field diffracted waves follow from eqns (171) and (134) and comparison with eqn (9) gives the angular dependence

$$A(\psi) = -\tfrac{1}{2} i\pi\, (Ka)^2 \tag{165}$$

so that from eqn (21) the mean drift force

$$\overline{X_1} = \tfrac{1}{4}\rho g \pi^2 a A^2\, (Ka)^3. \tag{166}$$

Note that this leading-order approximation to the drift force is independent of the submerged geometry of the structure, in particular it agrees with the calculation of McIver (1987, eqn 5.6) for the half-immersed sphere.

### 7.4 The half-immersed sphere

For a half-immersed sphere the boundary conditions (137) for the heave potential become

$$\frac{\partial \chi_3}{\partial R} = \cos\theta \quad \text{on} \quad R = a \quad \text{and} \quad \frac{\partial \chi_3}{\partial \theta} = 0 \quad \text{on} \quad \theta = \pi/2,\ r > a.$$

It is easily verified that a harmonic function satisfying these boundary conditions and with the asymptotic form (138) is

$$\chi_3 = -a \sum_{n=0}^{\infty} \frac{E_n P_{2n}(\cos\theta)}{(2n+1)(R/a)^{2n+1}} \tag{167}$$

where, from Hulme (1982, eqn B10),

$$\frac{E_n}{(4n+1)} = \int_0^1 P_{2n}(\nu) P_1(\nu)\, d\nu = \frac{(-1)^{n+1}(2n)!}{2^{2n+1}(n!)^2(2n-1)(n+1)} \tag{168}$$

and $P_m$ is a Legendre function of order $m$. The limiting added mass

$$\hat{a}_{33} = 2\rho\pi a^3 \sum_{n=0}^{\infty} \frac{E_n^2}{(2n+1)(4n+1)} = \frac{2\rho\pi a^3}{3} 0.830\,951\ldots \tag{169}$$

follows from eqn (142) where the numerical value for the series is given by Hulme (1982, eqn 3.18). By eqn (163), the heave exciting force

$$X_3 \sim -\rho g \pi a^2 A \left(1 - \tfrac{2}{3} Ka \times 1.830\,951\ldots\right). \tag{170}$$

The boundary conditions to be satisfied by the sway potential $\chi_1$ are

$$\frac{\partial \chi_1}{\partial R} = \sin\theta\cos\psi \quad \text{on} \quad R = a \quad \text{and} \quad \frac{\partial \chi_1}{\partial \theta} = 0 \quad \text{on} \quad \theta = \pi/2,\ r > a \tag{171}$$

and the solution with the far-field behaviour (147) is

$$\chi_1 = -\frac{a^3 \sin\theta\cos\psi}{2R^2} \tag{172}$$

so that the dipole coefficient $\mu_1 = 1/2$. From eqns (154), (157) and (162) the limiting value of the sway added mass is

$$\hat{a}_{11} = \rho\pi a^3/3 \tag{173}$$

and the asymptotic behaviours of the damping and exciting force are

$$b_{11} \sim \tfrac{1}{4}\rho\pi^2 a^3 \omega\,(Ka)^3 \tag{174}$$

and

$$X_1 \sim -i\rho g\pi a^2 A\,Ka \tag{175}$$

respectively.

## 8 Axisymmetric, submerged structures in deep water

The general comments made at the beginning of §5 apply here also. A compact, vertically axisymmetric structure of typical dimension $a$ is submerged to a depth $d$ in deep water. The wave length is assumed to be much smaller than $a$. Two cases may be distinguished depending on the relative magnitudes of $a$ and $d$. Davis & Leppington (1985) have considered scattering by a submerged sphere for the more difficult case $d/a = O(1)$ in the long wave regime as $Ka \to 0$. Here the dimension $a$ is taken to be much smaller than the other length scales in the problem so that both $a/d \ll 1$ and $Ka \ll 1$. No restriction is placed on the wavelength relative to the submergence so that $Kd$ is arbitrary.

The coordinate system is chosen so that the structure is placed symmetrically about $x = 0$ and the point $(x, z) = (0, d)$ is chosen as a reference point within the structure. Spherical polar coordinates $(R, \theta, \psi)$ are defined through

$$x = R\sin\theta\cos\psi, \quad y = R\sin\theta\sin\psi, \quad z = d + R\cos\theta. \tag{176}$$

### 8.1 Heave radiation problem

The free-surface no longer appears in the inner region, so the inner solution must satisfy only the body boundary condition and be able to match with the inner expansion of the outer solution. Thus, the leading order solution for the heave problem is written

$$\hat{\varphi}_3 = \chi_3 \tag{177}$$

where

$$\frac{\partial \chi_3}{\partial n} = n_3 \quad \text{on} \quad \Gamma \tag{178}$$

so that $\chi_3$ is the disturbance due to uniform motion of the structure in the $z$ direction. In the far field of the inner region

$$\chi_3 + \mu_3 a^3 (z-d)/R^3 \to 0 \quad \text{as} \quad R/a \to 0 \tag{179}$$

where $\mu_3$ is the dipole coefficient. To match with this the leading-order outer solution is a vertical wave dipole singular at $(x, y, z) = (0, 0, d)$, and so is written

$$\hat{\phi} = \alpha_3 G_{13}(x, y, z) \tag{180}$$

where (Thorne, 1953)

$$G_{13} = \frac{\cos\theta}{(KR)^2} - \frac{1}{K^2} \int_0^\infty \frac{\kappa + K}{\kappa - K} \kappa \, J_0(\kappa r) \, \mathrm{e}^{-\kappa(z+d)} \, d\kappa. \tag{181}$$

As $Kr \to \infty$

$$G_{13} \sim -2 \left( \frac{2\pi}{Kr} \right)^{1/2} \exp(-K(z+d) + i(Kr + \pi/4)) \tag{182}$$

and as $KR \to 0$

$$G_{13} = \frac{\cos\theta}{(KR)^2} + O(1). \tag{183}$$

Matching the inner and outer potentials (177) and (180), with the aid of the asymptotic forms (179) and (183), gives

$$\alpha_3 = -\mu_3 K^2 a^3. \tag{184}$$

The leading-order inner solution is real valued so eqn (17) yields only the limiting added mass

$$a_{33} \to \hat{a}_{33} = \int_\Gamma \chi_3 n_3 \, dS \quad \text{as} \quad a/d \to 0. \tag{185}$$

The added mass $\hat{a}_{33}$ is that of the structure moving uniformly through an infinite bath of fluid and is related to the dipole coefficient $\mu_3$ (Newman 1977, p. 143) by

$$\hat{a}_{33} = \rho(4\pi\mu_3 a^3 - V) \tag{186}$$

where $V$ is the volume of the structure.

The leading-order approximation to the damping may be determined form the amplitude of the waves radiated to the far field. Comparing the asymptotic form given by eqns (180) and (182) with (10) gives

$$\mathcal{A}_3(\psi) = -\frac{2\pi\mu_3 \, \mathrm{e}^{-Kd}}{\omega} (Ka)^3 \tag{187}$$

so that by eqn (19) the damping coefficient

$$b_{33} \sim \hat{b}_{33} = 8\rho\pi^2 a^3 \mu_3^2 \, \omega \, \mathrm{e}^{-2Kd} \, (Ka)^3 \quad \text{as} \quad a/d \to 0. \tag{188}$$

### 8.2 Sway radiation problem

Under the assumptions discussed at the beginning of §8, for a submerged body the sway radiation problem is very similar to the heave radiation problem but with the dipole oriented horizontally rather than vertically. The inner flow to leading order is

$$\hat{\varphi}_1 = \chi_1 \tag{189}$$

where $\chi_1$ is the disturbance resulting from uniform horizontal motion and satisfies

$$\frac{\partial \chi_1}{\partial n} = n_1 \quad \text{on} \quad \Gamma \tag{190}$$

with the far-field behaviour

$$\chi_1 + \mu_1 a^3 x/R^3 \to 0 \quad \text{as} \quad R/a \to \infty. \tag{191}$$

To leading order the outer solution is

$$\hat{\phi} = \alpha_1 G_{11}(x, y, z) \tag{192}$$

where the horizontal wave dipole (Thorne, 1953)

$$G_{11} = -\frac{\sin\theta\cos\psi}{(KR)^2} - \frac{\cos\psi}{K^2}\int_0^\infty \frac{\kappa + K}{\kappa - K}\kappa\, J_1(\kappa r)\,\mathrm{e}^{-\kappa(z+d)}\, d\kappa. \tag{193}$$

As $Kr \to \infty$

$$G_{11} \sim -2\left(\frac{2\pi}{Kr}\right)^{1/2} \exp(-K(z+d) + i(Kr - \pi/4))\cos\psi \tag{194}$$

and as $KR \to 0$

$$G_{11} = -\frac{\sin\theta\cos\psi}{(KR)^2} + O(KR). \tag{195}$$

Matching the inner and outer potentials (189) and (192) gives

$$\alpha_1 = \mu_1 K^2 a^3 \tag{196}$$

and hence, following the procedures used in §8.1, the limiting added mass is

$$a_{11} \to \hat{a}_{11} = \rho(4\pi\mu_1 a^3 - V) \tag{197}$$

and the damping coefficient

$$b_{11} \sim \hat{b}_{11} = 4\rho\pi^2 a^3 \mu_1^2 \omega\,(Ka)^3\,\mathrm{e}^{-2Kd}. \tag{198}$$

**8.3 Diffraction problem**

Within distances $O(a)$ of the structure, the incident wave

$$\phi_I = -\frac{igA}{\omega} e^{iKx} e^{-Kz} \sim -\frac{igA}{\omega} e^{-Kd}(1 + iK(x + i(z - d))) \tag{199}$$

so that the body boundary condition (6) may be approximated as

$$\frac{\partial \phi_D}{\partial n} = -\frac{\partial \phi_I}{\partial n} \sim -\omega A e^{-Kd}(n_1 + in_3). \tag{200}$$

As for the surface-piercing case described in §7.3, the inner solution to leading order for the diffraction potential combines the heave and sway radiation potentials so that

$$\hat{\varphi}_D = -\omega A e^{-Kd} \{\chi_1 + i\chi_3\}. \tag{201}$$

As each radiation potential is dipole like, matching is possible only if the leading-order outer potential also combines the two solutions for the radiation problems giving

$$\hat{\phi}_D = \omega A K^2 a^3 e^{-Kd} \{-\mu_1 G_{11} + i\mu_3 G_{13}\}. \tag{202}$$

The exciting forces follow from eqn (11) and are

$$X_1 \sim i\omega^2 \rho A e^{-Kd} \iint_\Gamma (x - \chi_1) n_1 \, dS = -\frac{igA}{a} Ka \, e^{-Kd}(\rho V + \hat{a}_{11}) \tag{203}$$

and

$$X_3 \sim i\omega^2 \rho A e^{-Kd} \iint_\Gamma i(z - d - \chi_3) n_3 \, dS = \frac{gA}{a} Ka \, e^{-Kd}(\rho V + \hat{a}_{33}). \tag{204}$$

Eqns (203) and (204) are the equivalent of the relative motion hypothesis for the case of a deeply submerged body, note the absence of a term involving the damping in these leading-order approximations. The exciting forces may also be expressed in terms of the dipole coefficients by use of eqns (186) and (197).

The far-field diffracted waves follow from eqns (202), (182) and (194) and comparison with eqn (9) gives the angular dependence

$$A(\psi) = 2i\pi \, (Ka)^3 \, e^{-2Kd}(\mu_1 \cos\psi + \mu_3) \tag{205}$$

so that from eqn (21) the mean drift force

$$\overline{X_1} = 2\rho g \pi^2 a A^2 \, (Ka)^5 \, e^{-4Kd} \left[(\mu_1 - \mu_3)^2 + \mu_3^2\right]. \tag{206}$$

### 8.4 The submerged sphere

For a sphere the dipole coefficients corresponding to uniform flows in the vertical and horizontal directions are each equal to a half. This immediately gives the leading-order approximations for a deeply submerged sphere

$$\hat{a}_{11} = \hat{a}_{33} = \tfrac{2}{3}\rho\pi a^3, \tag{207}$$

$$\hat{b}_{11} = \tfrac{1}{2}\hat{b}_{33} = \rho\pi^2 a^3 \omega\,(Ka)^3\ \mathrm{e}^{-2Kd}, \tag{208}$$

and

$$X_1 = -iX_3 = -2i\rho g\pi a^2 A\,Ka\ \mathrm{e}^{-Kd}. \tag{209}$$

Wang (1986) has given a very thorough analysis of the submerged sphere using multipole expansions and, in particular, he considered the true long-wave limit $Ka \to 0$ with $a$ and $d$ fixed. As $Kd \to 0$, the results for the added mass and damping in eqns (207) and (208) agree with the results of Wang if $a/d \to 0$.

## 9 Two-dimensional, surface-piercing structures in finite depth water

In this, and the following section the method of matched asymptotic expansions is used to obtain the hydrodynamic forces on isolated surface-piercing structures in finite depth water. This initial discussion is applicable to both the two- and three-dimensional cases.

All geometrical length scales are assumed to be of the same order of magnitude and will be typified by the depth $h$. The regime to be investigated is that of the wavelength being much greater than the other length scales in the problem so that

$$kh \ll 1. \tag{210}$$

As before, the flow domain is divided into two regions. The inner region surrounds the structure to horizontal distances much less than the wavelength so that $r \ll k^{-1}$ and includes a portion of the free surface and the horizontal bed. No wave effects are discernible in the inner region. The outer region is at horizontal distances much greater than the depth and is defined by $r \gg h$.

With the assumption of small $kh$, the free surface condition (7) may be rewritten as

$$\frac{\partial\phi}{\partial z} = -K\phi = -k\tanh kh\ \phi = -\frac{1}{h}\left[(kh)^2 - \tfrac{1}{3}(kh)^4 + \dots\right]\phi \quad \text{on} \quad z = 0 \tag{211}$$

where $\phi$ is any diffraction or radiation potential and eqn (4) has been used to replace $K$. This form of the free-surface condition suggests that both the inner and outer potentials may be expanded in $kh$ as

$$\phi = f(kh)\left[\hat{\phi}^{(0)} + kh\,\hat{\phi}^{(1)} + (kh)^2\,\hat{\phi}^{(2)} + \dots\right] \tag{212}$$

where $f(kh)$ is the order in $kh$ of the potential $\phi$ (see also Bai, 1977, §4.1). Inserting the expansion (212) into eqn (211) and equating like terms in $kh$ shows that $\hat{\phi}^{(0)}$ and $\hat{\phi}^{(1)}$ must each satisfy the homogeneous condition

$$\frac{\partial\hat{\phi}}{\partial z} = 0 \quad \text{on} \quad z = 0 \tag{213}$$

but the higher order terms must satisfy non-homogeneous conditions on the free surface.

The governing equation for $\phi$ within the fluid region is Laplace's equation and may be approximated by a simpler equation if the horizontal and vertical fluid motions are on different length scales in any part of the region. In the inner region there is no difference in the horizontal and vertical length scales. However, in the outer region, the horizontal motion takes place on the wavelength scale $k^{-1}$ and the vertical motion on the depth scale $h$. This can be made explicit by writing

$$\frac{\partial^2 \phi}{\partial x^2} + \frac{\partial^2 \phi}{\partial y^2} + \frac{\partial^2 \phi}{\partial z^2} = \frac{1}{h^2}\left[(kh)^2\left(\frac{\partial^2 \phi}{\partial (kx)^2} + \frac{\partial^2 \phi}{\partial (ky)^2}\right) + \frac{\partial^2 \phi}{\partial (z/h)^2}\right] = 0 \qquad (214)$$

where $kx$, $ky$ and $z/h$ derivatives are all of the same magnitude. So with an expansion of the form (212) each of the outer potentials $\hat{\phi}^{(0)}$ and $\hat{\phi}^{(1)}$ are solutions of

$$\frac{\partial^2 \hat{\phi}}{\partial z^2} = 0 \qquad (215)$$

which, when solved in conjunction with the free-surface condition (213) and bed condition (8), indicates that these two potentials must be independent of $z$. The outer solution must also satisfy the radiation condition specifying outgoing waves which leads to the solutions of the form

$$\hat{\phi} = \alpha\, e^{ik|x|} \qquad (216)$$

in two dimensions and

$$\hat{\phi} = \sum_{m=0}^{\infty} H_m(kr)\,(\alpha_m \cos m\psi + \beta_m \sin m\psi) \qquad (217)$$

in three dimensions, where $H_m$ denotes the Hankel function of the first kind and order $m$.

The two-dimensional case is discussed in the remainder of the present section. For simplicity all of the structures considered are assumed to be symmetric about the plane $x = 0$ and to intersect the mean free surface in $|x| \leq a$.

## 9.1 Heave radiation problem

Due to the fundamentally different behaviour of the added mass in the low frequency limit, in deep water it is logarithmically singular while in finite depth water it remains bounded, the heave radiation problem for a surface-piercing cylinder in finite-depth water has attracted considerable attention. For example, solutions have been obtained by Bai (1977), Kan (1970) and Sayer (1980). The solution of the heave radiation problem given here is based on that described in detail by McIver & Linton (1991). Unlike the problems considered previously, in the inner region it is necessary to consider terms other than that at leading order. A formal procedure is needed to match properly the inner and outer solutions but only an outline will be given and the reader interested in further details is referred to McIver & Linton.

Let $\chi_3$ be the harmonic function satisfying the body, bottom and free-surface boundary conditions

$$\frac{\partial \chi_3}{\partial n} = n_3 \quad \text{on} \quad \Gamma, \quad \frac{\partial \chi_3}{\partial z} = 0 \quad \text{on} \quad z = h \quad \text{and} \quad z = 0, \; |x| > a. \tag{218}$$

The body boundary condition on $\Gamma$ gives an instantaneous flow across the mean position of the structure and hence out of the inner region. At the outer limits of the inner region this will appear as a uniform stream and so the condition

$$\chi_3 - \delta_3 |x| \to 0 \quad \text{as} \quad |x/h| \to \infty \tag{219}$$

is imposed. Balancing the net flow into the inner region across $\Gamma$ with that exiting the inner region at large distances gives

$$-\int_\Gamma \frac{\partial \chi_3}{\partial n}\, dS = 2 \lim_{|x/h| \to \infty} \int_0^h \frac{\partial \chi_3}{\partial |x|}\, dz, \tag{220}$$

where the the result in eqn (62) is used to evaluate the integral on the left-hand side and the far-field form (219) is used on the right-hand side to obtain

$$\delta_3 = a/h. \tag{221}$$

McIver & Linton (1991) show that the leading-order solution in the inner region is not that at $O(1)$ generated directly by the body boundary condition. In fact, to match with the outer solution, it is required to include a lower-order constant term so that the inner solution

$$\hat{\varphi}_3 = \frac{\beta_3}{kh} + C_3 + \chi_3(x, z) \tag{222}$$

where $\beta_3$ and $C_3$ are $O(1)$ constants. As pointed out in the introduction to §9, the first two terms in the expansion in $kh$ of the inner solution satisfy the homogeneous free-surface condition (213) and hence so does $\hat{\varphi}_3$ as given in (222).

The leading-order outer solution has the form

$$\hat{\phi}_3 = \alpha_3 \, e^{ik|x|} \sim \alpha_3 (1 + ik|x|) \quad \text{as} \quad k|x| \to 0. \tag{223}$$

The outer expansion of $\hat{\varphi}_3$ follows from eqns (222) and (219) and matching the leading-order constant and the term in $x$ with (223) gives

$$\alpha_3 = \beta_3/kh \quad \text{and} \quad i\alpha_3 k = a/h \tag{224}$$

and hence

$$\alpha_3 = -ia/kh \quad \text{and} \quad \beta_3 = -ia. \tag{225}$$

The $O(1)$ constant, $C$, may be determined only by considering the inner solution up to $O(kh)$ and McIver & Linton (1991) show that

$$C_3 = -a^2/h. \tag{226}$$

The heave added mass and damping follow from eqn (17) and the inner solution (222) so that

$$a_{33} + i\frac{b_{33}}{\omega} = \rho\int_\Gamma \phi_3 n_3\, dS \sim \rho\int_\Gamma\left(-\frac{ia}{kh} - \frac{a^2}{h} + \chi_3\right) n_3\, dS. \tag{227}$$

Hence, as $kh \to 0$

$$a_{33} \sim \hat{a}_{33} = \rho\left(\frac{2a^3}{h} + \int_\Gamma \chi_3 n_3\, dS\right) \tag{228}$$

and

$$b_{33} \sim \hat{b}_{33} = \frac{2\rho a^2\omega}{kh}, \tag{229}$$

where the result in eqn (62) has been used to evaluate an integral. The damping coefficient may also be evaluated from eqn (27) using the far-field form of the radiation potential. Comparison of eqn (223), with $\alpha_3$ given by (225), with eqn (25) in the limit $kh \to 0$ gives

$$\mathcal{A}_3^+ = \mathcal{A}_3^- = \frac{a}{\omega h}kh \tag{230}$$

which when inserted in eqn (27) recovers (229).

McIver & Linton (1991) also considered the case of a submerged, heaving structure mounted on the bed and found that the low-frequency limit of the added mass is different from the above result. This is surprising as the low-frequency limiting boundary conditions for the surface and bottom-mounted cases are identical. However, the constant $C_3$ appearing in (222) is determined by the non-zero frequency behaviour. In the bottom-mounted case the constant is identically zero and so the limiting added mass does not contain the first term in (228).

For the case of a heaving rectangular block, an approximation to the limiting added mass may be found by using a 'thin gap' approximation, in which the clearance $d$ between the block and the bed is assumed to be much smaller than the block width $2a$. The heave potential is approximated by its depth averaged value, for the flow region beneath the block this is defined by

$$\chi_{3,d} = \frac{1}{d}\int_{h-d}^{h} \chi_3\, dz. \tag{231}$$

Integrating Laplaces equation over the gap and applying the bed and body conditions in eqns (218) gives

$$\frac{d^2\chi_{3,d}}{dx^2} = \frac{1}{d}, \quad |x| < a, \tag{232}$$

which has solution

$$\chi_{3,d} = \frac{x^2}{2d} + C_1, \tag{233}$$

where the symmetry has been used to eliminate a term in $x$. Similarly, for the region adjacent to the block, define the depth averaged potential

$$\chi_{3,h} = \frac{1}{h}\int_0^h \chi_3\, dz. \tag{234}$$

Integrating Laplace's equation over the depth and applying the surface and bed conditions in eqns (218) gives

$$\frac{d^2\chi_{3,h}}{dx^2} = 0, \quad |x| > a, \tag{235}$$

which has a solution

$$\chi_{3,h} = \frac{a}{h}|x| \tag{236}$$

that satisfies the condition (219). Continuity of pressure at the gap edge $|x| = a$ gives

$$a^2/2d + C_1 = a^2/h. \tag{237}$$

In the small gap approximation $\chi_3$ is replaced by $\chi_{3,d}$ when calculating the added mass, thus

$$\int_\Gamma \chi_3 n_3 \, dS \sim -\int_{-a}^{a} \chi_{3,d} \, dx = -2\int_0^a \left(\frac{x^2 - a^2}{2d} + \frac{a^2}{h}\right) dx = \frac{2a^3}{3d} - \frac{2a^3}{h}. \tag{238}$$

Thus, for a rectangular block with small clearance $d/a$ the limiting added mass

$$\hat{a}_{33} \sim \frac{2\rho a^3}{3d}. \tag{239}$$

As pointed out above, for a submerged block mounted on the bed the limiting added mass will not contain the contribution from the constant $C_3$ appearing in eqn (222) so that

$$\hat{a}_{33} \sim \frac{2\rho a^3}{3d}\left(1 - \frac{3d}{h}\right), \tag{240}$$

which will be negative if $d > h/3$. The phenomenon of negative added mass is well-known for heaving submerged structures (McIver & Evans, 1984) but this is the first geometry for which it has been found to occur at zero frequency. The above approximate solution assumes that the block is much wider than the gap width $d$, the phenomenon has been confirmed for a sufficiently wide rectangular block by accurate computation (C.M. Linton, private communication). An explicit solution for $\chi_3$ is possible for a thin plate on the bed and the limiting added mass is found to be negative provided that $a$ is greater than about $0.64h$.

## 9.2 Sway radiation problem

The sway radiation problem is more straightforward than the heave problem and the leading-order behaviour of the hydrodynamic coefficients can be determined by considering only the leading-order inner and outer solutions.

The leading-order inner solution is written

$$\hat{\varphi}_1 = \chi_1(x, z) \tag{241}$$

where $\chi_1$ satisfies the body boundary condition (6), the bed condition (8) and the leading-order approximation to the free-surface condition (211), that is

$$\frac{\partial \chi_1}{\partial n} = n_1 \quad \text{on} \quad \Gamma, \quad \frac{\partial \chi_1}{\partial z} = 0 \quad \text{on} \quad z = h \quad \text{and} \quad z = 0, \ |x| > a. \tag{242}$$

The potential $\chi_1$ is the disturbance to a uniform flow in the negative $x$ direction along a channel of width $h$ with a protrusion $\Gamma$ and in the far field

$$\chi_1 + L \operatorname{sgn} x \to 0 \quad \text{as} \quad |x/h| \to \infty. \tag{243}$$

The constant $L$ is known as the blockage coefficient and depends on the shape of $\Gamma$ and the depth $h$ (see Tuck, 1975).

The motion is antisymmetric in $x$ and so the leading order outer solution may be written

$$\hat{\phi}_1 = \operatorname{sgn} x \, \alpha_1 \, e^{ik|x|} \sim \operatorname{sgn} x \, \alpha_1 (1 + ik|x|) \quad \text{as} \quad k|x| \to 0. \tag{244}$$

Matching the asymptotic form in (244) with the outer expansion of the inner solution given by eqns (241) and (243) gives

$$\alpha_1 = -L \tag{245}$$

It follows from eqns (17) and (241) that the added mass

$$a_{11} \to \hat{a}_{11} = \rho \int_\Gamma \chi_1 n_1 \, dS \quad \text{as} \quad kh \to 0. \tag{246}$$

A number of authors have made calculations of this limiting added mass, for example Flagg & Newman (1971) have made calculations for rectangular cylinders. The limiting added mass is related to the blockage coefficient, an application of Green's theorem to $\chi_1$ and $\Psi \equiv x$ over the boundarys of the fluid domain with enclosing vertical lines at large $|x|$ gives

$$\hat{a}_{11} = \rho(2Lh - V) \tag{247}$$

where $V$ is the submerged cross-sectional area of the structure. The blockage coefficient for circular cylinders is discussed by Martin & Dalrymple (1988, appendix C).

The leading-order approximation to the damping coefficient follows from the far-field radiated waves. Comparing the outer solution (244) with eqn (25) as $kh \to 0$ gives

$$\mathcal{A}_1^+ = -\mathcal{A}_1^- = -\frac{i}{\omega}\frac{L}{h}(kh)^2 \tag{248}$$

so that from eqn (27)

$$b_{11} \sim \hat{b}_{11} = 2\rho L^2 \omega \, kh. \tag{249}$$

### 9.3 Diffraction problem

Within distances $O(h)$ of the structure the incident wave

$$\phi_I = -\frac{igA}{\omega \cosh kh} e^{ikx} \cosh k(z-h) \sim -\frac{igA}{\omega}\left(1 + ikx + \tfrac{1}{2}k^2\left[z^2 - 2zh - x^2\right]\right) \quad (250)$$

so the body boundary condition (6) for the diffracted field

$$\frac{\partial \phi_D}{\partial n} = -\frac{\partial \phi_I}{\partial n} \sim \frac{gkA}{\omega}\left[-\frac{\partial x}{\partial n} + \tfrac{1}{2}ik\frac{\partial}{\partial n}\left[z^2 - x^2\right] - ikh\frac{\partial z}{\partial n}\right]. \quad (251)$$

As with the cases considered previously, the low-frequency approximation to the diffraction potential is related to the sway and heave potentials (forced by the first and third terms on the right-hand side of eqn (251)). However, the two forcing terms have different orders of magnitude, the heave forcing term is multiplied by the small parameter $kh$, and there is a third term appearing. Here the aim is to obtain only the leading term in the inner potential. It has already been seen in §9.1 that the heave boundary condition forces a term at a lower order in $kh$ and so will make a contribution to the lowest order inner diffraction potential. A formal matching procedure can be used to show that the middle term on the right-hand will not affect the leading order term. From the results of §9.1 and §9.2, and in particular eqns (222) and (241), the leading-order inner potential is

$$\hat{\varphi}_D = -\frac{gkA}{\omega}(\chi_1 + a) \quad (252)$$

while, from eqns (223-5) and (243-4), the leading-order outer potential is

$$\phi_D = \frac{gkA}{\omega}(L \operatorname{sgn} x - a)\, e^{ik|x|}. \quad (253)$$

The wave forces follow from eqn (11). For consistency with the inner potential in eqn (251), the first two terms in the expansion of the incident wave in eqn (248) must be retained. The sway exciting force

$$X_1 = i\omega\rho \int_\Gamma \left(\frac{-igA}{\omega}\right)(ikx - ik\chi_1)n_1\, dS = -ikgA(\rho V + \hat{a}_{11}) \quad (254)$$

and the heave force is

$$\begin{aligned} X_3 &= i\omega\rho \int_\Gamma \left(\frac{-igA}{\omega}\right)(1 - ika)\, n_3\, dS \\ &= -2\rho gaA(1 - ika) = -2\rho gaA\left(1 - i\frac{(kh)^2}{2\rho ah}\frac{\hat{b}_{33}}{\omega}\right), \end{aligned} \quad (255)$$

which is the equivalent of the relative motion hypothesis for this case. This expression for the heave force cannot be used to obtain further terms in the damping expansion through the Haskind relation (28), as has been done in §5.3 for example, as the omitted terms in eqns (250-1) may contribute at the same order as $\hat{b}_{33}$.

Comparison of the outer potential (253) with eqns (23-4) as $kh \to 0$ gives the reflection and transmission coefficients

$$\mathcal{R} = -ik(L + a) \quad \text{and} \quad \mathcal{T} = 1 + ik(L - a) \quad (256)$$

respectively, which agree with Martin & Dalrymple (1988, eqn (127)).

## 10 Axisymmetric, surface-piercing structures in finite depth water

Here matched asymptotic expansions is used to investigate the radiation and scattering of low-frequency waves by an isolated structure in water of finite depth $h$. A discussion of the boundary value problem and its approximation for small $kh$, where $k$ is the wavenumber, is given at the beginning of §9. For simplicity the structures considered are assumed to be axisymmetric about the $z$ axis and to intersect the mean free surface in $r \leq a$.

### 10.1 Heave radiation problem

At low frequencies, the primary effect of a heaving structure in the free surface is to radiate waves axisymmetrically into the far field. The outer flow therefore appears as a wave source at the origin and so the leading-order outer solution is written

$$\hat{\phi}_3 = \alpha_3 H_0(kr) \tag{257}$$

where the constant $\alpha_3$ is to be found from matching and $H_m$ denotes the Hankel function of the first kind and order $m$. From Abramowitz & Stegun (1965),

$$H_0(kr) \sim \left(\frac{2}{\pi kr}\right)^{1/2} \mathrm{e}^{i(kr-\pi/4)} \quad \text{as} \quad kr \to \infty \tag{258}$$

and

$$H_0(kr) = 1 + \frac{2i}{\pi}\left(\gamma + \ln \tfrac{1}{2}kr\right) + O\left((kr)^2 \ln kr\right) \quad \text{as} \quad kr \to 0, \tag{259}$$

where $\gamma$ is Euler's constant.

Let $\chi_3$ be a harmonic function satisfying the bottom, body and free-surface conditions

$$\frac{\partial \chi_3}{\partial n} = n_3, \quad \frac{\partial \chi_3}{\partial z} = 0 \quad \text{on} \quad z = h \quad \text{and} \quad z = 0,\ r > a. \tag{260}$$

The body boundary condition generates an instantaneous net flow into the inner region and so the additional condition

$$\chi_3 - \alpha_3 \frac{2i}{\pi} \ln \frac{r}{h} \to 0 \quad \text{as} \quad \frac{r}{h} \to \infty \tag{261}$$

is imposed. The strength of the logarithm has been chosen to match that in the inner expansion of the outer solution given by eqn (257) with the asymptotic form (259). The constant $\alpha_3$ is found by equating the flow into the inner region,

$$\iint_\Gamma \left(-\frac{\partial \chi_3}{\partial n}\right) dS = -\iint_\Gamma n_3 \, dS = \pi a^2, \tag{262}$$

with the flow out of the inner region through a cylinder $S_\infty$ of large radius

$$\iint_{S_\infty} \frac{\partial \chi_3}{\partial r} dS = \lim_{r/h \to \infty} \int_0^h \left( \int_0^{2\pi} \alpha_3 \frac{2i}{\pi r} r \, d\psi \right) dz = 4ih\alpha_3 \tag{263}$$

to get

$$\alpha_3 = \pi a^2/4ih. \tag{264}$$

The inner expansion of the outer solution also contains constant terms (which satisfy homogeneous conditions on all boundaries of the inner region) so, to match fully with eqn (259), the complete leading-order approximation to the inner solution is written

$$\hat{\varphi}_3 = \alpha_3 \left[1 + \frac{2i}{\pi}\left(\gamma + \ln \tfrac{1}{2}kh\right)\right] + \chi_3. \tag{265}$$

The heave added mass and damping follow from eqns (17) and (265) and

$$a_{33} + i\frac{b_{33}}{\omega} = \rho \iint_\Gamma \phi_3 n_3 \, dS \sim \iint_\Gamma \left(\frac{\pi a^2}{4ih}\left[1 + \frac{2i}{\pi}\left(\gamma + \ln \tfrac{1}{2}kh\right)\right] + \chi_3\right) n_3 \, dS. \tag{266}$$

Hence, as $kh \to 0$,

$$a_{33} \sim \hat{a}_{33} = \rho\left(-\frac{\pi a^4}{2h}\left(\gamma + \ln \tfrac{1}{2}kh\right) + \iint_\Gamma \chi_3 \, dS\right) \tag{267}$$

and

$$b_{33} \sim \hat{b}_{33} = \frac{\rho \pi^2 a^4 \omega}{4h}. \tag{268}$$

Alternatively, the damping coefficient may be evaluated using the far field form of the radiation potential. Comparison of the outer potential (257), with the far field form given by (258), and eqn (10) with $kh \to 0$ gives

$$A_3(\psi) = \frac{\pi a^2}{4\omega h^2}(kh)^2 \tag{269}$$

which when substituted in eqn (18) recovers (266).

For the case of a heaving truncated vertical cylinder, Yeung (1981, p. 122) has found an approximation to the term in eqn (267) involving $\chi_3$ using a 'thin gap' approximation in which the clearance $d$ between the cylinder and the bed is assumed to be much smaller than the cylinder radius $a$. A depth averaged potential $\chi_{3,d}$ in the flow region beneath the block is defined as in eqn (231) and integrating Laplace's equation over the gap and applying the bed and body conditions in eqns (260) gives

$$\frac{1}{r}\frac{d}{dr}\left(r\frac{d\chi_{3,d}}{dr}\right) = \frac{1}{d} \tag{270}$$

which has solution

$$\chi_{3,d} = \frac{r^2}{4d} + C_1, \tag{271}$$

where a term singular at $r = 0$ has been excluded. Similarly, for the region adjacent to the block, define the depth averaged potential $\chi_{3,h}$ by eqn (234) and integrating Laplace's equation over the depth and applying the surface and bed conditions in eqns (260) then gives

$$\frac{1}{r}\frac{d}{dr}\left(r\frac{d\chi_{3,h}}{dr}\right) = 0 \tag{272}$$

which has a solution

$$\chi_{3,h} = \frac{a^2}{2h} \ln \frac{r}{h} \tag{273}$$

satisfying eqn (261). Continuity of pressure at the gap edge $r = a$ gives

$$\frac{a^2}{4d} + C_1 = \frac{a^2}{2h} \ln \frac{a}{h}. \tag{274}$$

In the small gap approximation $\chi_3$ is replaced by $\chi_{3,d}$ in calculating the added mass, thus

$$\iint_\Gamma \chi_3 n_3 \, dS \sim -\int_0^{2\pi} \left( \int_0^a \left( \frac{r^2 - a^2}{4d} + \frac{a^2}{2h} \ln \frac{a}{h} \right) r \, dr \right) d\psi = \frac{\pi a^4}{8d} - \frac{\pi a^4}{2h} \ln \frac{a}{h}. \tag{275}$$

So, for a truncated cylinder with small clearance $d/a$, the limiting added mass

$$\hat{a}_{33} \sim \frac{\rho \pi a^4}{2h} \left( \frac{h}{4d} - \gamma - \ln \tfrac{1}{2} ka \right), \tag{276}$$

as given by Yeung (1981, eqn 37).

### 10.2 Sway radiation problem

For the sway radiation problem the leading-order inner solution is written

$$\hat{\varphi}_1 = \chi_1 \tag{277}$$

where $\chi_1$ is the harmonic function that satisfies the body, bed and free-surface boundary conditions

$$\frac{\partial \chi_1}{\partial n} = n_1, \quad \frac{\partial \chi_1}{\partial z} = 0 \quad \text{on} \quad z = h \quad \text{and} \quad z = 0, \; r > a. \tag{278}$$

The potential $\chi_1$ is the disturbance due to uniform motion in the $x$ direction of the structure and its reflection in $z = 0$ and in the far field

$$\chi_1 + \mu_1 a^2 x / r^2 \to 0 \quad \text{as} \quad r/h \to \infty. \tag{279}$$

To match with the dipole behaviour in (279), the leading-order outer solution is written

$$\hat{\phi}_1 = \alpha_1 H_1(kr) \cos \psi \tag{280}$$

where, from Abramowitz & Stegun (1965),

$$H_1(kr) \sim -i \left( \frac{2}{\pi kr} \right)^{1/2} e^{i(kr - \pi/4)} \quad \text{as} \quad kr \to \infty \tag{281}$$

and

$$H_1(kr) = \frac{2}{\pi ikr} + O(kr \ln kr) \quad \text{as} \quad kr \to 0. \tag{282}$$

Matching the outer expansion of the inner solution, which follows from eqn (279), with the inner expansion of the outer solution, which follows from eqns (280) and (282), gives

$$\alpha_1 = -\tfrac{1}{2}\pi i\mu_1 ka^2. \tag{283}$$

From eqns (17) and (277), the added mass

$$a_{11} \to \hat{a}_{11} = \rho \iint_\Gamma \chi_1 n_1 \, dS \quad \text{as} \quad kh \to 0.$$

By an application of Green's theorem to $\chi_1$ and $\Psi = x$, the limiting added mass and the dipole coefficient are found to be related by

$$\hat{a}_{11} = \rho(2\pi\mu_1 a^2 h - V), \tag{284}$$

where $V$ is the submerged volume of the structure.

The leading-order approximation to the damping follows from the far-field form of the radiated waves. Comparing (10) with the far-field form of the outer solution given by eqns (280-1), yields

$$A_1(\psi) = -\frac{i\pi a^2 \mu_1}{2\omega h^2}(kh)^3 \cos\psi \tag{285}$$

and so from (19)

$$b_{11} \sim \hat{b}_{11} = \frac{\rho\pi^2 a^4 \mu_1^2 \omega}{2h}(kh)^2. \tag{286}$$

### 10.3 Diffraction problem

Within distances $O(h)$ of the structure the incident wave

$$\phi_I = -\frac{igA}{\omega \cosh kh} e^{ikx} \cosh k(z-h) \sim -\frac{igA}{\omega}\left(1 + ikx + \tfrac{1}{2}k^2\left[z^2 - 2zh - x^2\right]\right) \tag{287}$$

so the body boundary condition (6) for the diffracted field

$$\frac{\partial\phi_D}{\partial n} = -\frac{\partial\phi_I}{\partial n} \sim \frac{gkA}{\omega}\left[-\frac{\partial x}{\partial n} + \tfrac{1}{2}ik\frac{\partial}{\partial n}\left[z^2 - x^2\right] - ikh\frac{\partial z}{\partial n}\right]. \tag{288}$$

As with the cases considered previously, the low-frequency approximation to the diffraction potential is related to the sway and heave potentials forced by the first and third terms on the right-hand side of eqn (286). However, the situation is like that for the two-dimensional problem discussed in §9.3 in that the two forcing terms have different orders of magnitude and there is a third term appearing. Further, the third term does not contribute a net flow into the inner region and consequently cannot force any source terms in the outer solution.

A formal matching procedure may be used to show that only the sway and heave potentials in the inner solution contribute to the leading-order outer solution. The results of §10.1 and §10.2 now give the leading-order inner potential as

$$\hat{\varphi}_D = -\frac{gkA}{\omega}\chi_1 \tag{289}$$

and the leading-order outer potential as

$$\hat{\phi}_D = \frac{gkA}{\omega}\frac{\pi k a^2}{4}\left(-H_0(kr) + 2i\mu_1 H_1(kr)\cos\psi\right). \tag{290}$$

The leading-order to the sway exciting force is

$$\begin{aligned} \chi_1 &= i\omega\rho \iint_\Gamma \left(\frac{-igA}{\omega}\right)(ikx - ik\chi_1)n_1\, dS \\ &= -ikgA(\rho V + \hat{a}_{11}) = -2\pi i\rho g a h A \mu_1\, ka \end{aligned} \tag{291}$$

and to the heave exciting force is

$$\chi_3 = i\omega\rho \iint_\Gamma \left(\frac{-igA}{\omega}\right) n_3\, dS = -\rho g \pi a^2 A. \tag{292}$$

The approximation to the heave exciting force in eqn (292) is that due to the wave elevation at the structure. This has been used by Yeung (1981, p. 122) to obtain the leading-order behaviour of the added mass and damping. Substituting the result (292) into the Haskind relation (20) recovers the leading-order damping (268). The approximation (45) to one of the Kramers-Kronig relations then gives

$$a_{33} \sim -\frac{\hat{b}_{33}}{\pi\omega}\ln Ka = -\frac{\rho\pi a^4}{4h}\ln\left[(ka)^2\frac{h}{a}\right] \sim -\frac{\rho\pi a^4}{2h}\ln ka \quad \text{as} \quad ka \to 0, \tag{293}$$

in agreement with the dominant term in (267).

The far-field form of the diffraction potential (290) follows using eqn (258) and (281) and, by comparison with eqn (9), the angular dependence in the far field

$$\mathcal{A}(\psi) = -\tfrac{1}{4}\pi i(ka)^2(1 - 2\mu_1\cos\psi) \tag{294}$$

and hence, by (21), the drift force

$$\overline{X}_1 = \tfrac{1}{8}\rho g A^2 a\pi^2 (ka)^3\left(1 + 2\mu_1 + 2\mu_1^2\right). \tag{295}$$

For a circular cylinder extending throughout the depth, the dipole coefficient $\mu_1$ is unity recovering the result in McIver (1987, eqn 5.3) as $kh \to 0$.

## 11 Multi-element structures

The calculation of wave forces on structures with more than one element or, equivalently, the forces on a array of structures in close proximity to one another, is very much complicated by hydrodynamic interactions between the elements. Here, the available results for the low-frequency regime are reviewed.

As has been seen in the previous sections, the amplitude of the far-field waves radiated or scattered by an isolated structure are proportional to some power of the parameter $ka$, where $k$ is the wave number and $a$ is a typical structural dimension. In the limit $ka \to 0$ the far-field amplitudes approach zero so that it is tempting to conclude that for very low frequencies hydrodynamic interactions may be neglected when calculating wave forces. However, this argument is at fault as it neglects the importance of a third length scale in the problem, namely the typical element spacing $l$. For a fixed array geometry, the true low-frequency behaviour is obtained by letting the wave length become much greater than all other length scales in the problem so that, in particular $kl \ll 1$. In this regime wave effects are no longer discernible within the array and any interaction argument based on the far-field waves radiated or scattered by an isolated element is inappropriate.

It is possible to neglect hydrodynamic interactions to a first approximation if $ka \ll 1$ with $kl = O(1)$. This corresponds to a structure whose dimensions are much smaller than all other length scales in the problem.

Matched asymptotic expansions may be used to obtain information in the same way as described in §§5-10. That is, an outer region is defined that surrounds the array at horizontal radial distances $r \gg l$ and the outer solution will be one of the forms given previously. A second region may be defined within and surrounding the array to radial distances $r \ll k^{-1}$. However, although wave effects are not discernible within this region, it is still required to solve a complicated hydrodynamic interaction problem. Except for a few special cases, for example twin vertical cylinders for which bipolar coordinates may be used, this cannot be done without further approximation. McIver (1987, 1990) has made progress by making the further assumption that the structural elements are small compared to their spacing, that is

$$a \ll l \quad \text{as well as} \quad kl \ll 1. \tag{296}$$

Taken together these assumptions require $ka \ll 1$. With three different length scales, $k^{-1}$, $l$ and $a$, in the problem a scheme of matched asymptotic expansions requires the definition of three overlapping flow regions. An outer region is defined as above, from the point of view of an observer in the outer region the array appears as a single point disturbance. An intermediate region is defined within and surrounding the array to radial distances $r \ll k^{-1}$, but excluding the immediate vicinity of individual elements. An observer in this region will not observe any wave effects and will see each element as a point disturbance. Finally, inner regions are defined surrounding each structure to distances $r \ll l$. The solutions in each region turn out to be double expansions in the two small parameters $kl$ and $a/l$.

In the following two subsections a brief summary is given of the main results of the analysis by McIver (1987) of two array scattering problems involving circular cylinders and half-immersed spheres. There are no published calculations along the same lines for radiation problems. In the final subsection a brief outline is given of the heave radiation problem for axisymmetric, surface-piercing structures in finite depth water.

An array of $N$ elements is considered with the origin of coordinates chosen to be within the array. Each element is assumed to have a vertical axis of symmetry and the axis of element $j$ is located at $(x, y) = (x_j, y_j)$. Horizontal polar coordinates with origin at element $j$ are defined by

$$x - x_j = r_j \cos\psi_j, \quad y - y_j = r_j \sin\psi_j \tag{297}$$

while element $n$ is located at $(r_j, \psi_j) = (r_{jn}, \psi_{jn})$. Note that the relative polar coordinates satisfy the symmetry conditions

$$r_{nj} = r_{jn}, \quad \psi_{nj} = \psi_{jn} + \pi. \tag{298}$$

## 11.1 An array of vertical cylinders

Each structural element is taken to be a fixed vertical circular cylinder of radius $a$ and extending throughout the fluid depth $h$. The solution for scattering by the plane wave (3) will be described. With this geometry the depth dependence may be factored out of the problem by writing the total potential (incident plus diffracted)

$$\phi_I + \phi_D = -\frac{igA}{\omega}\left[e^{ikx} + \phi_S(x, y)\right]\frac{\cosh k(z-h)}{\cosh kh} \tag{299}$$

and so the solution is valid for arbitrary $kh$.

To leading-order in $kl$, the inner solution for cylinder $j$ (McIver, 1987, eqn A4) is

$$\begin{aligned}\hat{\varphi}_S = kl\,\frac{a}{l}\,i\Bigg[&\frac{\cos\psi_j}{r_j/a} - \frac{a}{l}\sum_{\substack{n=1\\n\neq j}}^{N}\frac{\cos\psi_{jn}}{r_{jn}/l}\\ &-\left(\frac{a}{l}\right)^2\sum_{\substack{n=1\\n\neq j}}^{N}\frac{\cos(\psi_j - 2\psi_{jn})}{r_{jn}^2/l^2}\left(\frac{r_j}{a} + \frac{a}{r_j}\right) + O\left((a/l)^3\right)\Bigg],\end{aligned} \tag{300}$$

where the sign of the second term has been corrected. Here, and elsewhere, the two small parameters are written explicitly wherever they occur. The first term on the right-hand side of (300) is the lowest order term in the direct response of element $j$ to the incident wave and corresponds to $\chi_1$ appearing in eqn (289). There will be a corresponding dipole in the inner region solution for each of the remaining $N-1$ array elements. It is the field generated by these other dipoles that forces the remaining terms on the right-hand side of eqn (300). The primary effect is a constant term, which is simply the sum of the dipoles evaluated at cylinder $j$. The secondary effect is to generate a uniform flow past cylinder $j$, and this is contained within the third term together with the dipole response required to give no flow through the cylinder.

If all angles are referred to the direction $\theta = \beta$, so that $\psi_j$ is replaced by $\psi_j - \beta$ etc, then eqn (300) is the inner potential for the waves scattered by a wave propagating in the direction $\theta = \beta$. The sway and surge components of the exciting force on cylinder $j$ follow from eqn (11) and are

$$X_{3j-2} = -2\pi i\rho g A a h \frac{\tanh kh}{kh} kl \frac{a}{l}\left[\cos\beta - \left(\frac{a}{l}\right)^2 \sum_{\substack{n=1\\n\neq j}}^{N} \frac{\cos(\beta - 2\psi_{jn})}{r_{jn}^2/l^2} + O\left((a/l)^4\right)\right] \tag{301}$$

and

$$X_{3j-1} = -2\pi i\rho g A a h \frac{\tanh kh}{kh} kl \frac{a}{l}\left[\sin\beta + \left(\frac{a}{l}\right)^2 \sum_{\substack{n=1\\n\neq j}}^{N} \frac{\sin(\beta - 2\psi_{jn})}{r_{jn}^2/l^2} + O\left((a/l)^4\right)\right] \tag{302}$$

respectively. The definition of the mode index in equation (1) has been used here. These expressions show that even the leading-order approximations to the exciting forces at low-frequencies are influenced by hydrodynamic interactions. In the limit $a/l \to 0$ and for $\beta = 0$, the sway force reduces to the single body result in eqn (291).

The introduction of the incidence angle $\beta$ into the exciting forces enables the Haskind relation (20) to be used to calculate damping coefficients. Direct use of eqns (301-2) gives damping coefficients corresponding to forced oscillations of one element while the remaining elements are held fixed. To obtain the total force on the structure it is only needed to sum over $j$. For example, if this is done for the sway exciting force in eqn (301) and the result (20) applied this gives the sway-sway damping as (McIver, 1987, eqn B10)

$$b_{11} = \frac{\rho\pi^2 a^2\omega^2 \tanh kh}{2k^2 c_g}(ka)^2\left[N^2 - 2\left(\frac{a}{l}\right)^2 N \sum_{j=1}^{N}\sum_{\substack{n=1\\n\neq j}}^{N} \frac{\cos 2\psi_{jn}}{r_{jn}^2/l^2} + O\left((a/l)^4\right)\right]. \tag{303}$$

Not only does the leading-order low-frequency behaviour involve hydrodynamic interaction terms, but the largest term increases by the square of the number of elements, even in the limit $a/l \to 0$. This $N^2$ enhancement of the damping has also been noted by Eatock-Taylor & Hung (1987).

To leading order in $kl$, the outer solution is

$$\begin{aligned}\hat{\phi}_S = (kl)^2\left(\frac{a}{l}\right)^2 \frac{i\pi}{4}\Bigg[-N H_0(kr) + 2i\, H_1(kr)\Bigg\{N\cos\psi \\ - \left(\frac{a}{l}\right)^2 \sum_{j=1}^{N}\sum_{\substack{n=1\\n\neq j}}^{N} \frac{\cos(\psi - 2\psi_{jn})}{r_{jn}^2/l^2} + O\left((a/l)^4\right)\Bigg\}\Bigg]\end{aligned} \tag{304}$$

which may be compared with the solution in eqn (290) for a single axisymmetric body in finite-depth water. The strength of the source term, involving $H_0(kr)$, is proportional to the number of elements but is not affected by hydrodynamic interactions. On the other hand the dipole strength has a term directly proportional to the number of elements but is also contains an interaction term. The far field form of $\hat{\phi}_S$ follows from (258) and (281) and comparison with (9) gives

$$\mathcal{A}(\psi) = (ka)^2 \frac{\pi i}{4}\left[N(2\cos\psi - 1) - 2\left(\frac{a}{l}\right)^2 \sum_{j=1}^{N}\sum_{\substack{n=1\\ n\neq j}}^{N} \frac{\cos(\psi - 2\psi_{jn})}{r_{jn}^2/l^2} + O\left((a/l)^4\right)\right]. \tag{305}$$

From eqn (21), the leading-order term in the low-frequency expansion of the in-line drift force is (McIver, 1987, eqn 5.3)

$$\overline{X}_1 = \tfrac{5}{8}\rho g A^2 a\pi^2 \frac{kc_g}{\omega}(ka)^3\left[N^2 - \tfrac{6}{5}N\left(\frac{a}{l}\right)^2 \sum_{j=1}^{N}\sum_{\substack{n=1\\ n\neq j}}^{N} \frac{\cos 2\psi_{jn}}{r_{jn}^2/l^2} + O\left((a/l)^4\right)\right]. \tag{306}$$

As with the damping coefficient in eqn (303), there is an $N^2$ enhancement of the drift force in a multi-element structure. This behaviour was first observed in numerical calculations by Eatock Taylor & Hung (1985). McIver (1987, §5) discusses the effects of the interaction term in eqn (306) for some specific geometries. The horizontal drift force perpendicular to the direction of wave advance follows from eqn (22) and is

$$\overline{X}_2 = -\tfrac{1}{4}\rho g A^2 a\pi^2 \frac{kc_g}{\omega}(ka)^3\left[N\left(\frac{a}{l}\right)^2 \sum_{j=1}^{N}\sum_{\substack{n=1\\ n\neq j}}^{N} \frac{\sin 2\psi_{jn}}{r_{jn}^2/l^2} + O\left((a/l)^4\right)\right]. \tag{307}$$

This force is zero whenever the structure is symmetric about the direction of wave advance.

## 11.2 An array of half-immersed spheres

Each element of the array is now taken to be half-immersed sphere of radius $a$. The water is taken to be infinitely deep. Here, spherical polar coordinates $(R_j, \theta_j, \psi_j)$ centred on sphere $j$, and defined in a similar way to those in eqn (131), are used.

To leading order in $Kl$, the inner solution for sphere $j$ is (McIver, 1987, eqn A8)

$$\begin{aligned}\hat{\varphi}_D = -\frac{igA}{\omega} Kl\,\frac{a}{l}\Bigg[&i\frac{\sin\theta_j\cos\psi_j}{2(R_j/a)^2} - \sum_{n=0}^{\infty}\frac{E_n P_{2n}(\cos\theta_j)}{(2n+1)(R_j/a)^{2n+1}} \\ &-\tfrac{1}{2}\sum_{\substack{n=1\\ n\neq j}}^{N}\Bigg\{\left(\frac{a}{l}\right)^2\frac{1}{r_{jn}/l} + \left(\frac{a}{l}\right)^3\Bigg(i\frac{\cos\psi_{jn}}{(r_{jn}/l)^2} \\ &+\frac{\cos(\psi_j - \psi_{jn})}{(r_{jn}/l)^2}\left(\frac{R_j}{a} + \frac{a^2}{2R_j^2}\right)\sin\theta_j\Bigg) + O\left((a/l)^4\right)\Bigg\}\Bigg]\end{aligned} \tag{308}$$

The first two terms here are the diffraction solution for a single sphere forced directly by the incident wave, and are given by eqns (160), (167) and (172). The first term is the dipole-like solution of a sway radiation problem and the second of a heave problem. The heave solution contains a simple source as its first term and it is these sources for the remaining $N-1$ spheres that generate the constant third term in eqn (308). The secondary effect of these $N-1$ sources is to generate a uniform flow past the sphere and this appears in the fifth term along with its dipole response. The fourth term is the leading order effect of the dipole terms in the inner regions of the other $N-1$ cylinders. The only source term that appears is that forced directly by the incident wave, no source terms are forced by the interactions between array elements.

The leading-order approximations to the sway, surge and heave exciting forces on sphere $j$ are respectively

$$X_{3j-2} = -\pi i \rho g A a^2 \, Ka \left[\cos\beta + \tfrac{1}{2} i \left(\frac{a}{l}\right)^2 \sum_{\substack{n=1 \\ n\neq j}}^{N} \frac{\cos\psi_{jn}}{(r_{jn}/l)^2} + O\left((a/l)^3\right)\right], \tag{309}$$

$$X_{3j-1} = -\pi i \rho g A a^2 \, Ka \left[\sin\beta - \tfrac{1}{2} i \left(\frac{a}{l}\right)^2 \sum_{\substack{n=1 \\ n\neq j}}^{N} \frac{\sin\psi_{jn}}{(r_{jn}/l)^2} + O\left((a/l)^3\right)\right] \tag{310}$$

and

$$X_{3j} = -\rho\pi g A a^2. \tag{311}$$

The horizontal forces are modified by hydrodynamic interactions but, to leading order, the heave force on any sphere is the same as if it were in isolation (see eqn (163)). The damping matrix may now be calculated from eqn (20). The total force on the array follows by summing over $j$ and (20) may also be used to calculate damping coefficients for the complete structure. In constrast to the vertical cylinders discussed in the previous section, the interaction terms in (309) and (311) do not contribute to the sway-sway and heave-heave coefficients and the results for the array are obtained by multiplying the single sphere results in eqns (144) and (174) by $N^2$.

To leading order in $Kl$, the outer solution is

$$\hat{\varphi}_D = \frac{igA}{2\omega} (Kl)^2 \left(\frac{a}{l}\right)^2 N G_0 \tag{312}$$

where $G_0$ is the wave source defined in eqn (133) and a sign has been corrected in McIver (1987, eqn (4.24)). Source terms can only be forced directly by the incident wave and so the total far-field potential is just $N$ times the result for a single element in eqn (161). Similarly, the far field angular dependence is

$$A(\psi) = -\tfrac{1}{2} i\pi N (Ka)^2 \tag{313}$$

and so, from (21), the in-line drift force is

$$\overline{X}_1 = \tfrac{1}{4}\rho g \pi^2 a A^2 N^2 (Ka)^3. \tag{314}$$

Again there is an $N^2$ enhancement of the drift force but, unlike the vertical cylinders, there are no terms in $a/l$ arising from hydrodynamic interactions.

### 11.3 An array of axisymmetric, surface-piercing structures in finite depth water

Although the full calculations for the radiation problems have not been carried out, it is not difficult to see that for the geometries considered in §§11.1-2 the low-frequency limits of the added mass coefficients for each array element will be the same as if the element were in isolation. Thus the total added mass is just the single element result multiplied by the number of elements. In the present section a geometry is discussed for which the leading-order term in the added mass at low frequency is proportional to the square of the number of elements.

Each element in the array is now considered to be an identical, surface-piercing structure with a vertical axis of symmetry. Each structure intersects the free surface in a disc of radius $a$ and the water has finite depth $h$. Results for isolated structures of this type are given in §10. Here the heave radiation problem is considered for the case when the whole array oscillates in phase. The results will be quoted without details of the derivation.

The form of the leading-order inner potential for each element is similar to that for the single structure case given in eqn (265). For element $j$ it is

$$\hat{\varphi}_3 = \frac{a^2}{2h}\left\{N\ln kl + \ln\frac{h}{l} + \frac{\pi N}{2i}\left(1 + \frac{2i}{\pi}(\gamma - \ln 2)\right) + \sum_{\substack{n=1\\ n\neq j}}^{N} \ln\frac{r_{nj}}{l}\right\} + \chi_3 \qquad (315)$$

where $\chi_3$ is the potential introduced in eqns (260-1) but now with the origin of coordinates at element $j$. For $N = 1$ this reduces to the single element case in eqn (265). Perhaps the most interesting aspect of eqn (315) is that there are contributions to the local pressure field that are proportional to the number of elements $N$ in the array, in contrast to the solutions in §§11.1-2. The leading-order outer solution is

$$\hat{\phi}_3 = \frac{\pi a^2 N}{4ih} H_0(kr) \qquad (316)$$

so that each heaving element contributes a wave source to the far field flow. The damping coefficient for the array follows from the far-field form of (316) and eqn (19) and is

$$b_{33} \sim \frac{\rho\pi^2 a^4 \omega N^2}{4h} \qquad (317)$$

which is proportional to $N^2$ as expected from the work of Eatock-Taylor & Hung (1987). The result (45) obtained from the Kramers-Kronig relations now gives the dominant term in the total added mass as

$$a_{33} \sim -\frac{\rho\pi a^4 N^2}{2h} \ln kl \qquad (318)$$

which again is proportional to the square of the number of array elements.

## 12. Conclusion

An extensive set of results has been given covering the low-frequency behaviour of hydrodynamic forces in many geometrical circumstances. There are a number of geometries for which the low-frequency asymptotic behaviour has not been investigated in detail. This includes submerged, isolated structures in finite depth water. For arrays of structural elements, radiation problems in particular have not been extensively investigated.

In addition to the work already mentioned, the method of matched asymptotic expansions has been used to obtain approximate solutions in a number of other circumstances related to low-frequency motions. For example, Eatock Taylor, Hu & Nielsen (1990) have made calculations of the drift force on a vertical cylinder advancing in waves,while McIver (1993) has considered the scattering of waves by a vertical cylinder in a narrow wave tank. There are no doubt many other problems for which the technique may usefully be employed.

## References

Abramowitz, M. & Stegun, I.A. (1965) *Handbook of Mathematical Functions.* Dover.

Bai, K.J. (1977) The added mass of two-dimensional cylinders heaving in water of finite depth. *J. Fluid Mechanics* **81**, 85-105.

Crighton, D.G. (1992) Matched asymptotic expansions applied to acoustics. In *Modern Methods in Analytical Acoustics: Lecture Notes*, 168-208. Springer-Verlag.

Davis, A.M.J. & Leppington, F.G. (1977) The scattering of electromagnetic surface waves by circular or elliptic cylinders. *Proceedings Royal Society of London* A.**353**, 55-75.

Davis, A.M.J. & Leppington, F.G. (1985) Scattering of long surface waves by circular cylinders or spheres. *Quarterly J. Mechanics and Applied Mathematics* **38**, 412-432.

Eatock Taylor, R. & Hung, S.M.(1985)Wave drift enhancement effects in multi-column structures. *Applied Ocean Research* **7**, 128-137.

Eatock Taylor, R. & Hung, S.M.(1987)Wave drift enhancement effects for semisubmersible and TLP systems. *J. Offshore Mechanics and Arctic Engineering* **109**, 327-334.

Eatock Taylor, R., Hu, S.U. & Nielsen, F.G. (1990) Mean drift forces on a slowly advancing vertical cylinder in long waves. *Applied Ocean Research* **12**, 141-152.

Flagg, C.N. & Newman, J.N. (1971) Sway added-mass coefficients for rectangular profiles in shallow water. *J. Ship Research* **15**, 257-265.

Gradshteyn, I.S. & Ryzhik, I.M. (1980) *Table of Integrals, Series, and Products.* Academic Press.

Greenhow, M. (1986) High- and low-frequency asymptotic consequences of the Kramers-Kronig relations. *J. Engineering Mathematics* **20**, 293-306.

Greenhow, M. & Ahn, S.I. (1988) Added mass and damping of horizontal circular cylinder sections. *Ocean Engineering* **15**, 495-504.

Hulme, A. (1982) The wave forces acting on a floating hemisphere undergoing forced periodic oscillations. *J. Fluid Mechanics* **121**, 443-463.

Kan, M. (1977) The added mass coefficient of a cylinder oscillating in shallow water in the limit $K \longrightarrow 0$ and $K \longrightarrow \infty$. *Papers of Ship Research Institute,* Tokyo, Japan. No. 52.

Kotik, J. & Mangulis, V. (1962) On the Kramers-Kronig relations for ship motions. *International Shipbuilding Progress* **9**, 3-10.

Leppington, F.G. (1992) Matched asymptotic methods in surface wave theory. In *Wave Asymptotics* (ed. P.A. Martin & G.R. Wickham), 69-85. Cambridge University Press.

Martin, P.A. & Dalrymple, R.A. (1988) Scattering of long waves by cylindrical obstacles and gratings using matched asymptotic expansions. *J. Fluid Mechanics* **188**,465-490.

McIver, P. (1987) Mean drift forces on arrays of bodies due to incident long waves. *J. Fluid Mechanics* **185**, 469-482.

McIver, P. (1990) The scattering of long water waves by a group of submerged horizontal cylinders. *Quarterly J. Mechanics and Applied Mathematics* **43**, 499-515.

McIver, P. (1993) The wave field scattered by a vertical cylinder in a narrow wave tank. *Applied Ocean Research* **15**, 25-37.

McIver, P. & Evans, D.V. (1984) The occurrence of negative added mass in free-surface problems involving submerged oscillating bodies. *J. Engineering Mathematics* **18**, 7-22.

McIver, P. & Linton, C.M. (1991) The added mass of bodies heaving at low frequency in water of finite depth. *Applied Ocean Research* **13**, 12-17.

Mei, C.C. (1983) *The Applied Dynamics of Ocean Surface Waves*. Wiley, New York.

Newman, J.N. (1977) *Marine Hydrodynamics*. MIT Press, Cambridge, Massachusetts.

Sayer, P. (1980) The long-wave behaviour of the virtual mass in water of finite depth. *Proceedings Royal Society of London* A.**372**, 65-91.

Simon, M.J. & Hulme, A. (1986) The radiation and scattering of long water waves. In *Hydrodynamics of Ocean Wave-Energy Utilization* (ed. D.V. Evans & A.F. de O. Falcão), 419-431. Springer-Verlag.

Thorne, R.C. (1953) Multipole expansions in the theory of surface waves. *Proceedings Cambridge Philosophical Society* **49**, 707-716.

Tuck, E.O. (1975) Matching problems involving flow through small holes. *Advances in Applied Mechanics* **15**, 89-158.

Ursell, F. (1950) Surface waves on deep water in the presence of a submerged cylinder. Parts I and II. *Proceedings Cambridge Philosophical Society* **46**, 141-158.

Ursell, F. (1964) The decay of the free motion of a floating body. *J. Fluid Mechanics* **19**, 305-319.

Wang, S. (1986) Motions of a spherical submarine in waves. *Ocean Engineering* **13**, 249-271.

Wehausen, J.V. (1971) The motion of floating bodies. *Annual Review of Fluid Mechanics* **3**, 237-268.

Yeung, R.W. (1981) Added mass and damping of a vertical cylinder in finite-depth waters. *Applied Ocean Research* **3**, 119-133.

# Chapter 2

# Second-order sum-frequency wave loads and response of a tension-leg platform

**T. Matsui**

*Department of Architecture, Nagoya University, Furo-cho, Chikusa-ku, Nagoya 464-01, Japan*

## Abstract

A complete second-order diffraction theory is developed for computing second-order sum-frequency wave loads on three-dimensional floating bodies. The theory is applied to predict the springing oscillation of a tension-leg platform. The theoretical predictions are compared with model test results. It is shown that the vertical-plane motions and resulting tether forces of a tension-leg platform include second-order sum-frequency resonant components, which are comparable in magnitude with first-order wave-frequency responses. The correlation between predicted and measured results is satisfactory, confirming the validity of the theory.

## Keywords

Second-order diffraction theory, nonlinear wave load, wave springing, sum-frequency oscillation, tension-leg platform, vertical-plane motion, tether force.

## 1 Introduction

The tension-leg platform (TLP) is now recognized as one of the most promising options for petroleum production in deep waters. With the TLP, the semi-submersible type structure is anchored by vertical tethers pre-tensioned through surplus buoyancy so that vertical-plane motions, e.g. heave, pitch and roll motions, are reduced substantially. The design of TLPs requires a number of additional design considerations associated with the dynamic behaviour of these platforms which have not been so critical for earlier shallow water platform designs. One of these considerations which is of vital concern and the subject of this chapter is the resonant high-frequency oscillations of the vertical tethers caused by higher order excitation forces.

The TLP is designed such that the natural frequencies of the vertical-plane motions (typically 1.5–3 rad/s) are substantially higher than the frequency band of the ocean waves. Also, the damping of the system is extremely low. Besides the wave frequency responses, the resonant vertical-plane oscillations and resulting tensions in the tethers, caused by any source of high-frequency excitations, then become critical for the design of the TLP. It is now widely accepted that the second-order sum-frequency wave exciting forces, often called 'springing' forces, are a primary source of such high-frequency resonant excitations. Thus, besides the need for better estimates of the damping forces, a reliable prediction of these nonlinear wave excitations is critical for the design of the TLP tethers against fatigue damage.

During the last fifteen years, considerable progress has been made on the development of the second-order wave diffraction theory and computations (Ogilvie, 1983). Despite such progresses, it is only recently that the complete calculations of the second-order sum-frequency wave excitations on a full TLP have been reported (Molin & Chen, 1990; Kim, 1991; Matsui, Suzuki & Sakoh, 1992). A primary reason is theoretical and computational difficulties in obtaining the solution of second-order sum-frequency potential. Thus, the previous calculations of springing forces on the TLP have been based on some simplifying approximations. For example, Yoshida *et al.* (1981) have taken into account only the second-order contribution due to change of wetted body surface along the waterline. Petrauskas & Liu (1987) have completely neglected the effects of second-order potential. De Boom, Pinkster & Tan (1984) have included the second-order incident potential contribution but neglected that of the diffraction counterparts. Consequently, large discrepancies were observed between predicted and measured results for the high-frequency wave excitations and responses of the TLP. For example, measured rms values of the tether forces reported by Petrauskas & Liu (1987) were three or more times larger than predictions using experimental damping values.

The complete calculation of the nonlinear sum-frequency wave excitations on a TLP was first attempted by Kim & Yue (1988). Based on their complete second-order diffraction theory for axisymmetric bodies (Kim & Yue, 1989, 1990), they obtained the sum-frequency wave loads on a single TLP column and applied those results to the dynamic analysis of a full TLP through a simple approximation method (e.g. multiplying the single-leg results by the number of columns to estimate the wave loads on the full TLP). The computed rms values of the tether force were found to be two to three times higher than that predicted by an approximate second-order theory using second-order contributions due to first-order potentials only. However, the wave interaction between columns and the pontoon effects are ignored in their estimation.

The full three-dimensional analyses of the second-order sum-frequency wave loads on a fixed TLP have been reported more recently by Molin & Chen (1990), Kim (1991) and Matsui, Suzuki & Sakoh (1992). Through comparison of the full TLP results with approximate evaluations based on the results of a single column (Kim and Yue, 1988) and four columns only (Chau, 1989), Kim (1991) has shown that both the wave interaction between columns and the effect of pontoons are important to the calculation of the springing forces on the TLP. Similar calculations have been presented by Molin & Chen (1990).

Very recently, Matsui, Sakoh & Nozu (1992, 1993) have conducted model tests to measure the second-order sum-frequency wave excitations, and resultant motions and tether forces of a full TLP. These test results were compared with the predictions based on the complete second-order diffraction theory (Matsui, Suzuki & Sakoh, 1992). Satisfactory agreement was observed between computed and measured results, promising the usefulness of the complete second-order diffraction theory for predicting the sum-frequency wave excitations and responses of the TLP.

In this chapter, the complete second-order diffraction theory is developed and applied to predict the nonlinear sum-frequency wave loads and response of a full TLP. In Section 2, the boundary-value problems for the second-order diffraction theory are formulated based on the potential flow assumptions. Green's second identity is exploited to evaluate the second-order wave forces without the need for explicitly obtaining the solution of the second-order sum-frequency potential. The resulting expressions for the second-order forces contain the free-surface integral and the double spatial gradients of the first-order potentials, which are generally difficult to evaluate. The methodologies are discussed in some detail for the effective evaluation of these quantities. To illustrate the validity of the numerical procedure, a number of examples are studied for bodies of simple geometry. In Section 3, the equations of motion for the TLP undergoing high-frequency oscillations are derived, taking into account the geometric nonlinearities of the tether system (e.g. set-down effects due to horizontal platform motions). It is shown that the second-order contributions due to these nonlinearities provide additional forcing functions exciting the high-frequency motions of the TLP. In order to validate the theoretical predictions, model tests have been performed, as outlined in Section 4. In Section 5, the test results are presented and compared with the theoretical predictions. The conclusions are summarised in the final section.

## 2 Second-order diffraction theory for floating bodies

### 2.1 Statement of boundary-value problem

Consider the first- and second-order diffraction of plane regular incident waves by a three-dimensional floating body. The fluid is assumed to be invicid, incompressible and irrotational. The motion of the body may be characterised by the six degrees-of-freedom: surge ($\Xi_1$), sway ($\Xi_2$), heave ($\Xi_3$), roll ($\Xi_4$), pitch ($\Xi_5$) and yaw ($\Xi_6$). A space-fixed Cartesian co-ordinate system $oxyz$ is defined as shown in Fig. 1 with the $oxy$-plane on the mean free surface and the $oz$-axis measured positive vertically upwards.

A point on the body is denoted by the position vector $\mathbf{x}$ relative to the body-fixed co-ordinate axes with an origin at the centre of gravity of the body. The displacement vector $\mathbf{X}$ of the point $\mathbf{x}$, referred to the space-fixed co-ordinate axes, may then be expressed as

$$\mathbf{X} = \mathbf{U} + (\underline{D} - \underline{I})\mathbf{x} \tag{1}$$

where

$$\mathbf{U} = \Xi_1\mathbf{e}_1 + \Xi_2\mathbf{e}_2 + \Xi_3\mathbf{e}_3 \tag{2}$$

is the displacement vector of the centre of gravity, $\mathbf{e}_j (j = 1, 2, 3)$ denoting the unit

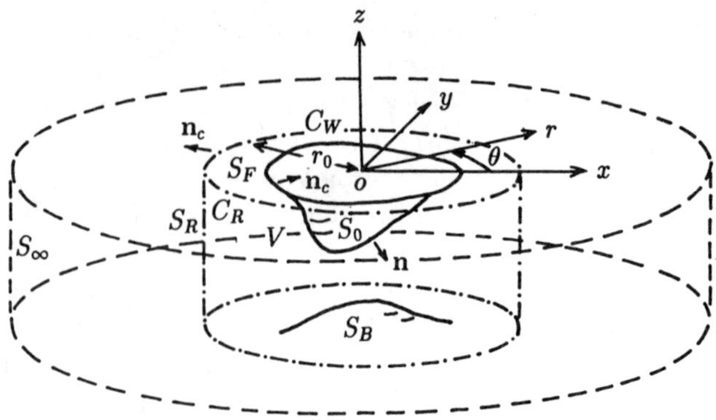


Figure 1: Co-ordinate system and fluid boundaries.

vectors in the direction of the space-fixed co-ordinate axes,

$$\underline{D} = \begin{bmatrix} \cos\Xi_5\cos\Xi_6 & -\cos\Xi_5\sin\Xi_6 & \sin\Xi_5 \\ \cos\Xi_4\sin\Xi_6+\sin\Xi_4\sin\Xi_5\cos\Xi_6 & \cos\Xi_4\cos\Xi_6-\sin\Xi_4\sin\Xi_5\sin\Xi_6 & -\sin\Xi_4\cos\Xi_5 \\ \sin\Xi_4\sin\Xi_6-\cos\Xi_4\sin\Xi_5\cos\Xi_6 & \sin\Xi_4\cos\Xi_6+\cos\Xi_4\sin\Xi_5\sin\Xi_6 & \cos\Xi_4\cos\Xi_5 \end{bmatrix} \tag{3}$$

and $\underline{I}$ is the $3\times 3$ unit matrix. Equation (3) implies that roll, pitch and yaw are taken in that order (Ogilvie, 1983).

Based on the assumption of small amplitude motions, it is reasonable to use the perturbation expansion

$$\mathbf{X} = \varepsilon\mathbf{X}^{(1)} + \varepsilon^2\mathbf{X}^{(2)} + O(\varepsilon^3) \tag{4}$$

for $\mathbf{X}$, where $\varepsilon$ is the usual perturbation parameter proportional to the wave slope. The first- and second-order motion vectors $\mathbf{X}^{(1)}$, $\mathbf{X}^{(2)}$ may then be written as

$$\mathbf{X}^{(1)} = \mathbf{U}^{(1)} + \mathbf{W}^{(1)}\times\mathbf{x} \tag{5}$$

$$\mathbf{X}^{(2)} = \mathbf{U}^{(2)} + \mathbf{W}^{(2)}\times\mathbf{x} + \underline{H}^{(2)}\mathbf{x}_m \tag{6}$$

where

$$\mathbf{U}^{(n)} = \Xi_1^{(n)}\mathbf{e}_1 + \Xi_2^{(n)}\mathbf{e}_2 + \Xi_3^{(n)}\mathbf{e}_3, \qquad n=1,2 \tag{7}$$

are the first- and second-order displacement vectors of the centre of gravity,

$$\mathbf{W}^{(n)} = \Xi_4^{(n)}\mathbf{e}_1 + \Xi_5^{(n)}\mathbf{e}_2 + \Xi_6^{(n)}\mathbf{e}_3, \qquad n=1,2 \tag{8}$$

are the first- and second-order small rotation vectors, and

$$\underline{H}^{(2)} = -\frac{1}{2}\begin{bmatrix} \Xi_5^{(1)2}+\Xi_6^{(1)2} & 0 & 0 \\ -2\Xi_4^{(1)}\Xi_5^{(1)} & \Xi_4^{(1)2}+\Xi_6^{(1)2} & 0 \\ -2\Xi_4^{(1)}\Xi_6^{(1)} & -2\Xi_5^{(1)}\Xi_6^{(1)} & \Xi_4^{(1)2}+\Xi_5^{(1)2} \end{bmatrix} \tag{9}$$

In a regular wave of frequency $\omega$, the first-order body motions may be expressed as

$$\mathbf{X}^{(1)} = Re\left[\mathbf{x}^{(1)}\mathrm{e}^{-i\omega t}\right] \tag{10}$$

$$\mathbf{U}^{(1)} = Re\left[\mathbf{u}^{(1)}\mathrm{e}^{-i\omega t}\right] \tag{11}$$

$$\mathbf{W}^{(1)} = Re\left[\mathbf{w}^{(1)}\mathrm{e}^{-i\omega t}\right] \tag{12}$$

$$\Xi_k^{(1)} = Re\left[\xi_k^{(1)}\mathrm{e}^{-i\omega t}\right], \qquad k = 1, 2\cdots, 6 \tag{13}$$

at time $t$. The second-order body motions can be decomposed into a mean and time harmonic component at twice the wave frequency. In the formulation described below, these second-order motions are assumed to be completely restraint. This implies that the expressions derived for the second-order forces correspond to the wave exciting forces. (The second-order forces induced by the second-order body motions can be obtained in the manner analogous to that in the first-order radiation analysis. This will be discussed in Section 3.)

With the assumption of ideal and irrotational flow, the fluid motion is defined by a velocity potential $\Phi$, which can be expressed in the form of the Stokes expansion

$$\Phi = \varepsilon\Phi^{(1)} + \varepsilon^2\Phi^{(2)} + O(\varepsilon^3) \tag{14}$$

The first-order velocity potential for a regular wave of frequency $\omega$ is

$$\Phi^{(1)} = Re\left[\phi^{(1)}\mathrm{e}^{-i\omega t}\right] \tag{15}$$

which can be decomposed into incident and diffraction components

$$\Phi^{(1)} = \Phi_I^{(1)} + \Phi_D^{(1)} = Re\left[(\phi_I^{(1)} + \phi_D^{(1)})\mathrm{e}^{-i\omega t}\right] \tag{16}$$

For an incident wave of amplitude $\zeta$ travelling in the direction $\theta = \alpha$, the first-order complex incident potential is

$$\phi_I^{(1)} = -\frac{ig\zeta}{\omega}\frac{\cosh k(z+h)}{\cosh kh}\mathrm{e}^{ik(x\cos\alpha + y\sin\alpha)} \tag{17}$$

where $g$ is the acceleration due to gravity, $h$ is the water depth, and $k$ is the wave number that satisfies

$$k\tanh kh = \frac{\omega^2}{g} \equiv \nu \tag{18}$$

The problem of the first-order diffraction potential is classical, and a variety of numerical methods of solution is now available.

The associated second-order velocity potential is defined by

$$\Phi^{(2)} = \Phi_I^{(2)} + \Phi_D^{(2)} \tag{19}$$

which fulfills (Ogilvie, 1983)

$$\nabla^2\Phi^{(2)} = 0 \tag{20a}$$

in the fluid domain $V$,

$$\nabla\Phi^{(2)} \cdot \mathbf{n} = 0 \tag{20b}$$

on the seabed $S_B$,

$$g\frac{\partial\Phi^{(2)}}{\partial z} + \frac{\partial^2\Phi^{(2)}}{\partial t^2} = -2\nabla\Phi^{(1)} \cdot \nabla\left(\frac{\partial\Phi^{(1)}}{\partial t}\right) + \frac{\partial}{\partial z}\left(\frac{\partial\Phi^{(1)}}{\partial z} + \frac{1}{g}\frac{\partial^2\Phi^{(1)}}{\partial t^2}\right)\frac{\partial\Phi^{(1)}}{\partial t} \tag{20c}$$

on the free surface $S_F$, and

$$\nabla\Phi^{(2)} \cdot \mathbf{n} = -\left[(\mathbf{X}^{(1)} \cdot \nabla)\nabla\Phi^{(1)}\right] \cdot \mathbf{n} + \left(\frac{\partial \mathbf{X}^{(1)}}{\partial t} - \nabla\Phi^{(1)}\right) \cdot (\mathbf{W}^{(1)} \times \mathbf{n})$$
$$+ \left(\frac{\partial \underline{H}^{(2)}}{\partial t}\mathbf{x}\right) \cdot \mathbf{n} \tag{20d}$$

on the body surface $S_0$, where $\mathbf{n}$ is the normal vector directed outwardly from the body, referenced to the body axes.

The second-order incident potential may be obtained by substituting eqn (17) into the right-hand side of eqn (20c), yielding

$$\Phi_I^{(2)} = Re\left[\phi_I^{(2)} \mathrm{e}^{-i2\omega t} + \delta_I^{(2)} t\right] \tag{21}$$

where

$$\phi_I^{(2)} = \frac{i3\omega\zeta^2}{8}\frac{\cosh 2k(z+h)}{\sinh^4 kh}\mathrm{e}^{i2k(x\cos\alpha + y\sin\alpha)} \tag{22}$$

The constant $\delta_I^{(2)}$ is determined from the requirement that the mean water level be $z = 0$ in an otherwise undisturbed incident wave, giving

$$\delta_I^{(2)} = -\frac{gk\zeta^2}{2\sinh 2kh} \tag{23}$$

The analogue expression for the second-order diffraction potential is

$$\Phi_D^{(2)} = Re\left[\phi_D^{(2)} \mathrm{e}^{-i2\omega t}\right] \tag{24}$$

By substituting eqns (10)–(13), (15), (19) and (21) into eqns (20a–d), the equations governing the second-order diffraction potential $\phi_D^{(2)}$ read

$$\nabla^2\phi_D^{(2)} = 0 \qquad \text{in } V \tag{25a}$$

$$\nabla\phi_D^{(2)} \cdot \mathbf{n} = 0 \qquad \text{on } S_B \tag{25b}$$

$$\frac{\partial\phi_D^{(2)}}{\partial z} - \frac{4\omega^2}{g}\phi_D^{(2)} = \alpha^{(2)} \qquad \text{on } S_F \tag{25c}$$

$$\nabla\phi_D^{(2)} \cdot \mathbf{n} = -\nabla\phi_I^{(2)} \cdot \mathbf{n} + \beta^{(2)} \qquad \text{on } S_0 \tag{25d}$$

where

$$\alpha^{(2)} = \frac{i\omega}{g}\left[\nabla\phi^{(1)}\cdot\nabla\phi^{(1)} - \nabla\phi_I^{(1)}\cdot\nabla\phi_I^{(1)} - \frac{1}{2}\phi^{(1)}\left(\frac{\partial^2\phi^{(1)}}{\partial z^2} - k^2\tanh^2 kh\phi^{(1)}\right)\right.$$
$$\left.+\frac{1}{2}\phi_I^{(1)}\left(\frac{\partial^2\phi_I^{(1)}}{\partial z^2} - k^2\tanh^2 kh\phi_I^{(1)}\right)\right] \tag{26}$$

$$\beta^{(2)} = -\frac{1}{2}\left[(\mathbf{x}^{(1)}\cdot\nabla)\nabla\phi^{(1)}\right]\cdot\mathbf{n} - \frac{1}{2}(\mathbf{w}^{(1)}\times\mathbf{n})\cdot(i\omega\mathbf{x}^{(1)} + \nabla\phi^{(1)})$$
$$+\frac{i\omega}{2}(\underline{h}^{(2)}\mathbf{x})\cdot\mathbf{n} \tag{27}$$

with

$$\underline{h}^{(2)} = \begin{bmatrix} \xi_5^{(1)2} + \xi_6^{(1)2} & 0 & 0 \\ -2\xi_4^{(1)}\xi_5^{(1)} & \xi_4^{(1)2} + \xi_6^{(1)2} & 0 \\ -2\xi_4^{(1)}\xi_6^{(1)} & -2\xi_5^{(1)}\xi_6^{(1)} & \xi_4^{(1)2} + \xi_5^{(1)2} \end{bmatrix} \tag{28}$$

In addition, $\phi_D^{(2)}$ must satisfy the radiation boundary condition ensuring the outgoing wave at infinity. It is shown later that the second-order forces can be evaluated without the need for explicitly solving the boundary-value problem (25).

## 2.2 Expressions for second-order forces

The wave forces exerted on the body can be obtained by integrating the hydrodynamic pressure over the submerged body surface. By retaining terms up to second order in the integrated pressure, the first- and second-order forces on vertically wall-sided bodies are found to be (Ogilvie, 1983)

$$\mathbf{F}^{(1)} = \rho\iint_{S_0}\frac{\partial\Phi^{(1)}}{\partial t}\mathbf{n}dS \tag{29}$$

$$\mathbf{F}^{(2)} = \rho\iint_{S_0}\left[\frac{\partial\Phi^{(2)}}{\partial t} + \frac{1}{2}\mid\nabla\Phi^{(1)}\mid^2 + \mathbf{X}^{(1)}\cdot\nabla\left(\frac{\partial\Phi^{(1)}}{\partial t}\right)\right]\mathbf{n}dS$$
$$-\frac{\rho g}{2}\oint_{C_W}\zeta_R^{(1)^2}\mathbf{n}dl + \mathbf{W}^{(1)}\times\left(m\frac{d^2\mathbf{U}^{(1)}}{dt^2}\right)$$
$$-\frac{\rho g}{2}A_W(x_f\Xi_4^{(1)} + y_f\Xi_5^{(1)})\Xi_6^{(1)}\mathbf{e}_3 \tag{30}$$

where $\zeta_R^{(1)}$ is the first-order relative wave elevation defined by

$$\zeta_R^{(1)} = -\frac{1}{g}\frac{\partial\Phi^{(1)}}{\partial t} - \mathbf{X}^{(1)}\cdot\mathbf{e}_3 = Re\left[\eta_R^{(1)}e^{-i\omega t}\right] \tag{31}$$

along the mean waterline $C_W$, $\rho$ is the fluid density, $m$ is the mass of body, $A_W$ is the water plane area, and $(x_f, y_f)$ are the co-ordinates of the centre of flotation.

The terms in eqn (30), depending on quadratic products of the first-order effects, can be decomposed into a mean ($\mathbf{F}_m^{(2)}$) and time oscillatory component ($\mathbf{F}_o^{(2)}$)

respectively, such that

$$\mathbf{F}_m^{(2)} = Re\left[\frac{\rho}{2}\iint_{S_0}\left(\frac{1}{2}\nabla\phi^{(1)}\cdot\nabla\phi^{(1)*} - i\omega\nabla\phi^{(1)}\cdot\mathbf{x}^{(1)*}\right)\mathbf{n}dS \right.$$
$$-\frac{\rho g}{4}\oint_{C_W}\eta_R^{(1)}\eta_R^{(1)*}\mathbf{n}dl - \frac{m\omega^2}{2}\mathbf{w}^{(1)}\times\mathbf{u}^{(1)*}$$
$$\left. -\frac{\rho g}{4}A_W(x_f\xi_4^{(1)} + y_f\xi_5^{(1)})\xi_6^{(1)*}\mathbf{e}_3\right] \tag{32}$$

$$\mathbf{F}_o^{(2)} = Re\left\{\left[\frac{\rho}{2}\iint_{S_0}\left(\frac{1}{2}\nabla\phi^{(1)}\cdot\nabla\phi^{(1)} - i\omega\nabla\phi^{(1)}\cdot\mathbf{x}^{(1)}\right)\mathbf{n}dS \right.\right.$$
$$-\frac{\rho g}{4}\oint_{C_W}\eta_R^{(1)^2}\mathbf{n}dl - \frac{m\omega^2}{2}\mathbf{w}^{(1)}\times\mathbf{u}^{(1)}$$
$$\left.\left. -\frac{\rho g}{4}A_W(x_f\xi_4^{(1)} + y_f\xi_5^{(1)})\xi_6^{(1)}\mathbf{e}_3\right]e^{-i2\omega t}\right\} \tag{33}$$

where $*$ denotes the complex conjugate. The calculation of $\mathbf{F}_m^{(2)}$ and $\mathbf{F}_o^{(2)}$ is straightforward for any body for which the first-order solution is known.

The remaining term in eqn (30), namely

$$\mathbf{F}_q^{(2)} = \rho\iint_{S_0}\frac{\partial\Phi^{(2)}}{\partial t}\mathbf{n}dS \tag{34}$$

can be evaluated in the manner suggested by Lighthill (1979) and Molin (1979). To do this, use is made of the auxiliary first-order radiation potential $\psi_j$, associated with a unit oscillation of the body in the direction $\mathbf{e}_j$ at double frequency $2\omega$. By applying Green's second identity, and making use of eqns (25a–d) and radiation boundary condition for $\phi_D^{(2)}$, $\mathbf{F}_q^{(2)}$ can be expressed in terms of the first-order quantities alone, such that

$$\mathbf{F}_q^{(2)} = Re\left[(\mathbf{F}_{qI} + \mathbf{F}_{qII} + \mathbf{F}_{qIII} + \mathbf{F}_{qIV})e^{-i2\omega t}\right] + \rho A_W\delta_I^{(2)}\mathbf{e}_3 \tag{35}$$

where

$$\mathbf{F}_{qI} = -2i\rho\omega\iint_{S_0}\phi_I^{(2)}\mathbf{n}dS \tag{36}$$

$$\mathbf{F}_{qII} = \sum_{j=1}^{3}\left(2i\rho\omega\iint_{S_0}\psi_j\nabla\phi_I^{(2)}\cdot\mathbf{n}dS\right)\mathbf{e}_j \tag{37}$$

$$\mathbf{F}_{qIII} = \sum_{j=1}^{3}\left(-2i\rho\omega\iint_{S_0}\psi_j\beta^{(2)}dS\right)\mathbf{e}_j \tag{38}$$

$$\mathbf{F}_{qIV} = \sum_{j=1}^{3}\left(2i\rho\omega\iint_{S_F}\psi_j\alpha^{(2)}dS\right)\mathbf{e}_j \tag{39}$$

The calculation of $\mathbf{F}_{qI}$ and $\mathbf{F}_{qII}$ is straightforward, since $\phi_I^{(2)}$ is known and $\psi_j$ can be obtained from the first-order solution. The difficulty arises in the evaluation of $\mathbf{F}_{qIII}$ containing the double spatial gradients of the first-order potentials, for which accurate solutions are difficult to obtain by existing numerical methods. Moreover, in order to evaluate $\mathbf{F}_{qIV}$, the integration must be performed over an infinite area of the free surface, but this is not trivial because the integrand displays a highly oscillatory and slowly decaying behaviour.

### 2.3 Evaluation of double spatial gradients of first-order potentials

The double spatial gradients of the first-order potentials involved in $\mathbf{F}_{qIII}$ can be reduced to single gradients by virture of Stokes' theorem

$$\iint_{S_0} \nabla \times (\psi_j \mathbf{x}^{(1)} \times \nabla\phi^{(1)}) \cdot \mathbf{n} dS = \oint_{C_W} (\psi_j \mathbf{x}^{(1)} \times \nabla\phi^{(1)}) \cdot d\mathbf{l} \tag{40}$$

where the vector $d\mathbf{l}$ is directed as indicated in Fig. 2. Using the well-known formulae of vector analysis, and by virtue of the assumption of irrotational and ideal flow, it follows that (Ogilvie & Tuck, 1969)

$$\begin{aligned} &\iint_{S_0} \psi_j \left[(\mathbf{x}^{(1)} \cdot \nabla)\nabla\phi^{(1)}\right] \cdot \mathbf{n} dS \\ &= \iint_{S_0} \Big[(\mathbf{x}^{(1)} \cdot \mathbf{n})(\nabla\psi_j \cdot \nabla\phi^{(1)}) - (\mathbf{x}^{(1)} \cdot \nabla\psi_j)(\mathbf{n} \cdot \nabla\phi^{(1)}) \\ &\quad -\psi_j(\mathbf{w}^{(1)} \times \mathbf{n}) \cdot \nabla\phi^{(1)}\Big] dS - \oint_{C_W} (\psi_j \mathbf{x}^{(1)} \times \nabla\phi^{(1)}) \cdot d\mathbf{l} \end{aligned} \tag{41}$$

which allows the double spatial gradients of the potential $\phi^{(1)}$ to be replaced by single gradients of $\phi^{(1)}$ and the auxiliary potential $\psi_j$.

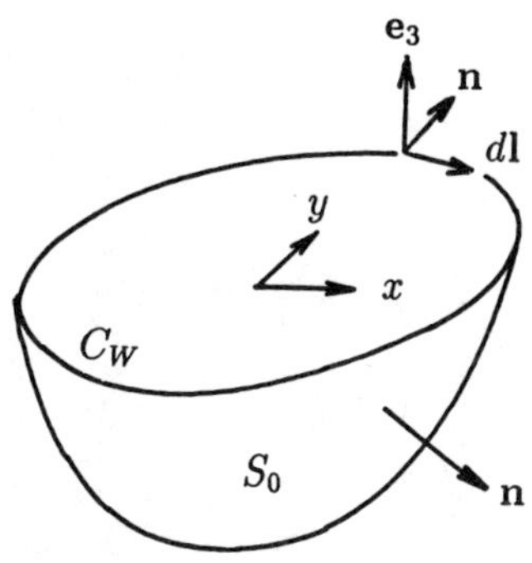


Figure 2: Definition of geometry for Stokes' theorem.

For wall-sided bodies that intersect the free surface at right angles, the vector $d\mathbf{l}$ is horizontal and defined by

$$d\mathbf{l} = (\mathbf{n} \times \mathbf{e}_3) dl \tag{42}$$

The line integral in the right-hand side of eqn (41) can then be given by

$$\oint_{C_W} (\psi_j \mathbf{x}^{(1)} \times \nabla\phi^{(1)}) \cdot d\mathbf{l} = \oint_{C_W} \psi_j \left[(\mathbf{x}^{(1)} \cdot \mathbf{n})\frac{\partial\phi^{(1)}}{\partial z} - (\mathbf{x}^{(1)} \cdot \mathbf{e}_3)\frac{\partial\phi^{(1)}}{\partial n}\right] dl \tag{43}$$

Combining eqns (27), (41) and (43) with eqn (38), $\mathbf{F}_{q\mathrm{III}}$ can be rewritten in the form

$$\mathbf{F}_{q\mathrm{III}} = \sum_{j=1}^{3}\Bigg\{ i\rho\omega \iint_{S_0} \Big[ (\mathbf{x}^{(1)}\cdot\mathbf{n})(\nabla\psi_j\cdot\nabla\phi^{(1)}) - (\mathbf{x}^{(1)}\cdot\nabla\psi_j)(\mathbf{n}\cdot\nabla\phi^{(1)}) + i\omega\psi_j(\mathbf{w}^{(1)}\times\mathbf{n})\cdot\mathbf{x}^{(1)} + \frac{i\omega}{2}(\underline{h}^{(2)}\mathbf{x})\cdot\mathbf{n}\Big] dS - i\rho\omega \oint_{C_W} \psi_j \left[ (\mathbf{x}^{(1)}\cdot\mathbf{n})\frac{\partial\phi^{(1)}}{\partial z} - (\mathbf{x}^{(1)}\cdot\mathbf{e}_3)\frac{\partial\phi^{(1)}}{\partial n} \right] dl \Bigg\} \tag{44}$$

which is more desirable for numerical evaluation.

## 2.4 Evaluation of free-surface integral

For the evaluation of the free-surface integral of eqn (39), it is convenient to separate the free surface $S_F$ into the discretised domain $S_{F1}$ close to the body and the infinite outer domain $S_{F2}$. The integral over the discretised domain can then be evaluated by numerical quadrature, while it is handled asymptotically for the infinite domain.

The free-surface integral consists of a summation of integrals of the general form

$$\iint_{S_F} \psi_j D_1\phi_1^{(1)} D_2\phi_2^{(1)} dS \tag{45}$$

where $D_1$ and $D_2$ are differential operators defined by eqn (26). Either one or both of the potentials $\phi_1^{(1)}$ and $\phi_2^{(1)}$ are diffraction potentials. These integrals in the discretised domain can be evaluated accurately by numerical quadrature (e.g. a four-point Gaussian quadrature) if the integrand contains only the single spatial gradients of the potentials.

In order to evaluate the integrals involving the double spatial gradients, it is useful to reduce the double derivatives to single derivatives by virtue of Green's theorem in two-dimensions. This yields

$$\begin{aligned}\iint_{S_{F1}} \psi_j\phi_1^{(1)}\frac{\partial^2\phi_2^{(1)}}{\partial z^2} dS &= -\iint_{S_{F1}} \psi_j\phi_1^{(1)}\left(\frac{\partial^2\phi_2^{(1)}}{\partial x^2} + \frac{\partial^2\phi_2^{(1)}}{\partial y^2}\right) dS \\ &= \iint_{S_{F1}} \left[\frac{\partial(\psi_j\phi_1^{(1)})}{\partial x}\frac{\partial\phi_2^{(1)}}{\partial x} + \frac{\partial(\psi_j\phi_1^{(1)})}{\partial y}\frac{\partial\phi_2^{(1)}}{\partial y}\right] dS \\ &\quad - \oint_{C_W+C_R} \psi_j\phi_1^{(1)}\frac{\partial\phi_2^{(1)}}{\partial n_c} dl\end{aligned} \tag{46}$$

where $C_R$ is a fictitious circular boundary separating the free surface into the two domains, and $n_c$ denotes the normal directed outwardly from $C_W$ and $C_R$ (refer to Fig. 1). Since the integrand of eqn (46) contains only single derivatives of $\phi^{(1)}$ and $\psi_j$, the integral can be evaluated accurately from solutions based on conventional numerical first-order analysis.

The integrals (45) over the infinite outer domain may be evaluated by means of an asymptotic method. For this purpose it is convenient to introduce polar co-ordinates $(r, \theta)$, such that

$$x = r\cos\theta, \qquad y = r\sin\theta \tag{47}$$

The technique is discussed below for the special case

$$I = \iint_{S_{F2}} \psi_j \phi_D^{(1)2} dS \tag{48}$$

For the sake of compactness, the wave flow is assumed to be symmetric with respect to $\theta = 0$.

Away from the body the potentials may be written as

$$\phi_D^{(1)} = \sum_{n=0}^{\infty} \varepsilon_n \left[ A_n(z) H_n^{(1)}(kr) + \sum_{m=1}^{\infty} A_{nm}(z) K_n(k_m r) \right] \cos n\theta = \phi_D^H + \phi_D^K \tag{49}$$

$$\psi_j = \sum_{n=0}^{\infty} \varepsilon_n \left[ \alpha_n(z) H_n^{(1)}(\kappa r) + \sum_{m=1}^{\infty} \alpha_{nm}(z) K_n(\kappa_m r) \right] \cos n\theta = \psi^H + \psi^K \tag{50}$$

where $\varepsilon_n = 1$ for $n = 0$, otherwise $\varepsilon_n = 2$, $H_n^{(1)}$ is the Hankel function of the first kind of order $n$, $K_n$ is the modified Bessel function of the second kind of order $n$, and $k_m$, $\kappa$ and $\kappa_m$ denote the real positive roots of the equations

$$k_m \tan k_m h = -\omega^2/g \tag{51}$$

$$\kappa \tanh \kappa h = 4\omega^2/g \tag{52}$$

$$\kappa_m \tan \kappa_m h = -4\omega^2/g \tag{53}$$

The second terms of eqns (49) and (50), $\phi_D^K$, $\psi^K$, represent local (evanescent) modes that decay exponentially with the radial distance, and the far-field asymptotics of the potentials are given only by the first terms, $\phi_D^H$, $\psi^H$, which represent propagating waves.

The integral (48) may now be evaluated over two intervals, $(r_0, r_1)$ and $(r_1, \infty)$, such that

$$I = \int_0^{2\pi} d\theta \left\{ \int_{r_0}^{r_1} \psi_j \phi_D^{(1)2} r dr + \int_{r_1}^{\infty} \psi^H (\phi_D^H)^2 r dr + \int_{r_1}^{\infty} \left[ \psi_j \phi_D^{(1)2} - \psi^H (\phi_D^H)^2 \right] r dr \right\} \tag{54}$$

where $r_0$ is the radius of the fictitious boundary $C_R$, and the radius $r_1$ is chosen so that the influence of local modes vanishes in the latter interval. The first integral in eqn (54) over the finite interval $(r_0, r_1)$ can be computed numerically. The last integral is made negligibly small by a suitable choice of the partition radius $r_1$.

The second integral over the infinite domain

$$I_2 = \int_0^{2\pi} d\theta \int_{r_1}^{\infty} \psi^H (\phi_D^H)^2 r dr \tag{55}$$

may be evaluated in the following manner. The integral is first executed over the azimuthal angle $\theta$. By substituting eqns (49) and (50) into eqn (55), and integrating over $\theta$ using the method of stationary phase, one obtains

$$I_2 = \pi \sum_{n=0}^{\infty} \sum_{m=0}^{\infty} \varepsilon_n \varepsilon_m A_n(0) \left[ \alpha_{|n-m|}(0) I(n, m, | n - m |) + \alpha_{n+m}(0) I(n, m, n + m) \right] \tag{56}$$

where

$$I(n,m,l) = \int_{r_1}^{\infty} H_n^{(1)}(kr)H_m^{(1)}(kr)H_l^{(1)}(\kappa r)r\,dr \tag{57}$$

At large radius one may use the large argument asymptotic form of Hankel function (Abramowitz & Stegun, 1972)

$$H_n^{(1)}(x) \sim \left(\frac{2}{\pi x}\right)^{\frac{1}{2}} h_n(x)\mathrm{e}^{i(x-\chi_n)} \tag{58}$$

where

$$h_n(x) = 1 + \sum_{s=1}^{\infty} i^s \frac{(4n^2-1^2)(4n^2-3^2)\cdots[4n^2-(2s-1)^2]}{s!(8x)^{2s}} \tag{59}$$

and $\chi_n = (2n+1)\pi/4$. When this is substituted into eqn (57), one obtains

$$I(n,m,l) = \int_{r_1}^{\infty} F(r)\mathrm{e}^{iqr}\,dr \tag{60}$$

where

$$F(r) = \frac{2}{\pi k}\left(\frac{2}{\pi\kappa r}\right)^{\frac{1}{2}} \mathrm{e}^{-i\chi_{n+m+l}} h_n(kr)h_m(kr)h_l(\kappa r) \tag{61}$$

$$q = 2k + \kappa \tag{62}$$

By virtue of the periodic nature of the exponential function of pure complex argument, the infinite integral (60) can be replaced by a finite-interval integral

$$I(n,m,l) = \int_{r_1}^{r_1+\frac{\pi}{q}} G(r)\mathrm{e}^{iqr}\,dr \tag{63}$$

where

$$G(r) = \sum_{s=0}^{\infty}(-1)^s F\left(r + \frac{s\pi}{q}\right) \tag{64}$$

Since $F(r)$ decreases as $r^{-1/2}$, it is difficult to obtain a converged solution by direct summation of the series (64). The convergence may be improved if the series is replaced by Euler's transformation (Abramowitz & Stegun, 1972)

$$\sum_{s=0}^{\infty}(-1)^s F_s = \sum_{s=0}^{\infty}(-1)^s \frac{\Delta^s F_0}{2^{s+1}} \tag{65}$$

where

$$F_s = F\left(r + \frac{s\pi}{q}\right) \tag{66}$$

and $\Delta^s F_0$ is the $s$th difference of $F_0$ given by

$$\Delta^s F_0 = \sum_{t=0}^{\infty}(-1)^t \binom{s}{t} F_{s-t} \tag{67}$$

One may note that the convergence rate of the series (65) is more rapid for $F_s$ and that it decays more slowly with increasing $s$. The convergence may further be improved if

Euler's transformation is applied after the first few (say $M$) terms of the series (64) are directly summed, i.e.

$$\sum_{s=0}^{\infty}(-1)^s F_s = \sum_{s=0}^{M-1}(-1)^s F_s + \sum_{s=0}^{\infty}(-1)^{s+M}\frac{\Delta^s F_M}{2^{s+1}} \tag{68}$$

The finite integral of eqn (63) can then be evaluated to any desirable degree of accuracy by applying numerical quadrature (e.g. a Newton-Cotes 9-point rule).

## 2.5 Application to bodies of simple geometry

The theory described in the foregoing was applied to develop a general numerical procedure for computing the second-order wave exciting forces on a three-dimensional floating body of arbitrary form. The expressions derived for the second-order forces were evaluated from numerical first-order solutions based on the hybrid integral-equation method, described by Matsui and Kato (1991). Figure 3 illustrates typical idealisations for the hybrid integral-equation method. The fluid domain was divided into two regions by introducing a fictitious vertical cylinder enclosing the body. The boundary surfaces inside the fictitious cylinder were discretised into quadratic isoparametric elements with 8 nodes, while an analytical solution based on the eigenfunction expansion in cylindrical co-ordinates was employed for the infinite outer domain.

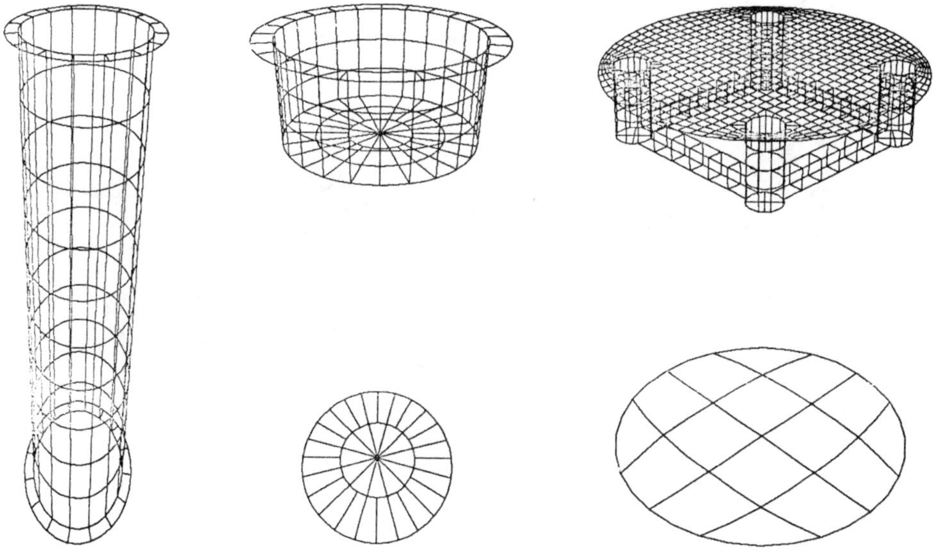

Figure 3: Boundary element idealisations for uniform vertical cylinder (left), truncated vertical cylinder (middle), and four-column TLP (right).

In order to illustrate the validity of the present numerical procedure, a number of numerical examples have been studied for bodies of simple geometry.

The geometry studied first is a vertical circular cylinder fixed at the seabed and piercing the free surface. For this case an analytical solution for the second-order

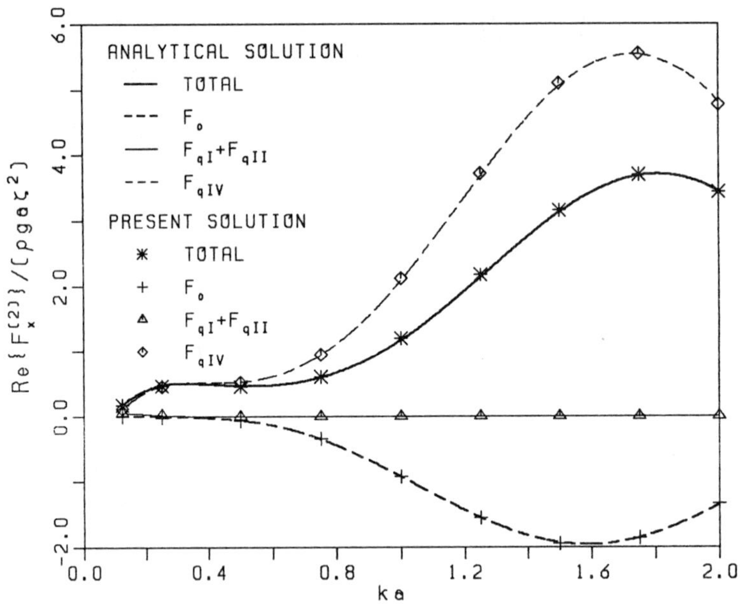


(a) in-phase component

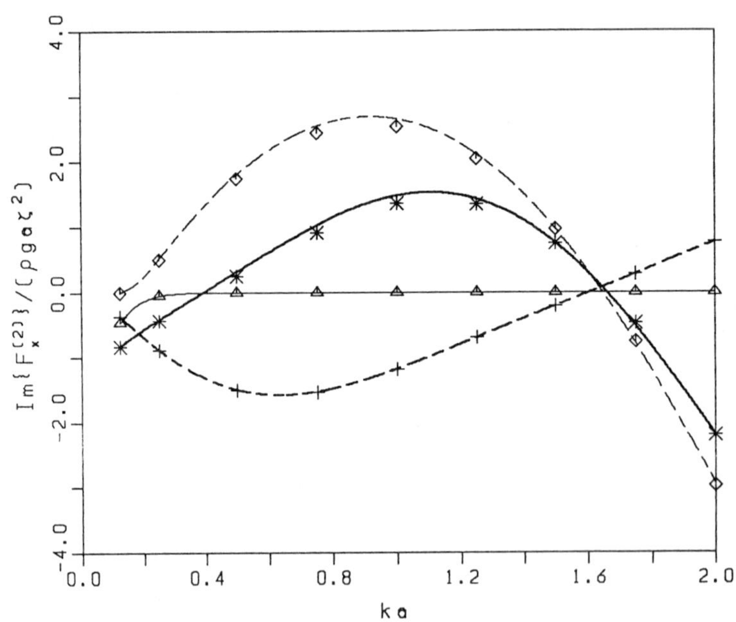


(b) out-of-phase component

Figure 4: Comparison between analytical and numerical results for second-order double-frequency horizontal force on fixed uniform vertical cylinder.

Table 1: Contribution of free-surface integral to second-order double-frequency horizontal force on uniform vertical cylinder ($r_0/a$=1.2, $ka$=1.0)

| $r_1/a$ | Local wave effects included | Local wave effects excluded |
|---|---|---|
| 3 | $3.425+2.406i$ | $4.236+1.564i$ |
| 4 | $3.228+2.538i$ | $4.236+1.564i$ |
| 5 | $3.128+2.546i$ | $4.236+1.564i$ |
| 6 | $3.069+2.570i$ | $4.236+1.564i$ |
| 8 | $2.997+2.596i$ | $4.236+1.564i$ |
| 10 | $2.960+2.607i$ | $4.236+1.564i$ |
| 12 | $2.938+2.614i$ | $4.236+1.564i$ |
| 16 | $2.916+2.620i$ | $4.236+1.564i$ |
| 20 | $2.907+2.623i$ | $4.236+1.564i$ |

double-frequency horizontal force has been obtained by several authors (e.g. Molin, 1979; Rahman, 1984; Eatock Taylor & Hung, 1887), and serves as a useful basis for checking the validity of the numerical solution. Based on the results of systematic convergence tests (Matsui, Suzuki & Sakoh, 1992), the double series of eigenfunction expansion were truncated at 9th Fourier harmonics in $\theta$ and at 14 evanescent modes in $z$. The double symmetry of the body geometry was exploited. Figure 4 shows the computed results for the double-frequency horizontal force on the cylinder, which are compared with the corresponding analytical solutions. The depth-to-radius radio of the cylinder is assumed to be $h/a = 10$. Agreement between analytical and numerical predictions is seen to be excellent over the frequency range examined, confirming the validity and accuracy of the present numerical procedure.

As an example illustrating the effectiveness of the procedure for evaluating the free-surface integral discussed above, Table 1 gives the results for the contribution of the free-surface integral to the double-frequency horizontal force on the cylinder. Both the results including and excluding the local wave effects are presented with increasing the partition radius $r_1$. The fact that the integral excluding the local wave effects gives a stationary value, independent of $r_1$, demonstrates the effectiveness and high accuracy of the proposed method. As the local wave effects decay exponentially with the radial distance, the integral including these effects approaches quickly to the asymptotic value with increasing $r_1$.

To give experimental evidence on the validity of the present numerical procedure, the second example studied is a truncated vertical cylinder fixed near the free surface, for which model test results have been presented by Matsui, Suzuki & Sakoh (1992). Figures 5 and 6 show the comparison between computed and measured second-order horizontal and vertical forces on the cylinder having a draught-to-radius ratio of $d/a = 1$. Both the results computed with and without the second-order potential effects are presented and compared with the test results. It is observed that the contribution of the second-order potential is dominant over the entire frequency range.

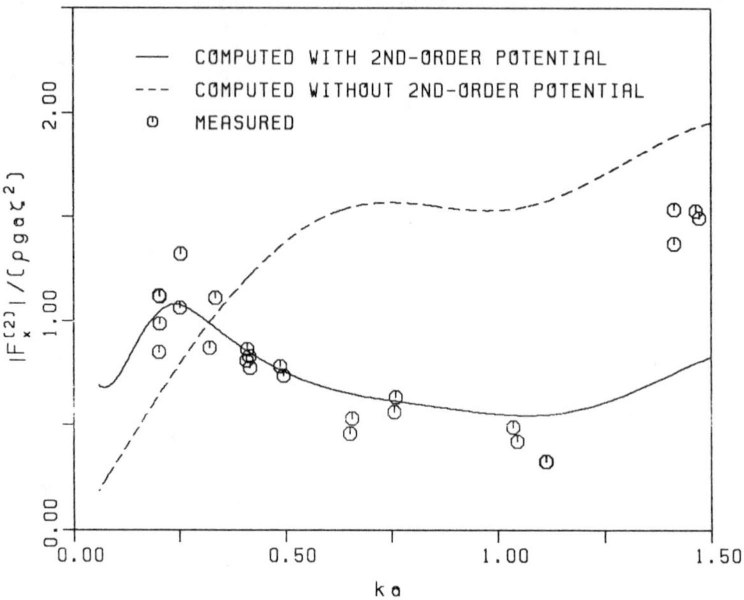


Figure 5: Comparison between computed and measured second-order double-frequency horizontal forces on fixed truncated vertical cylinder.

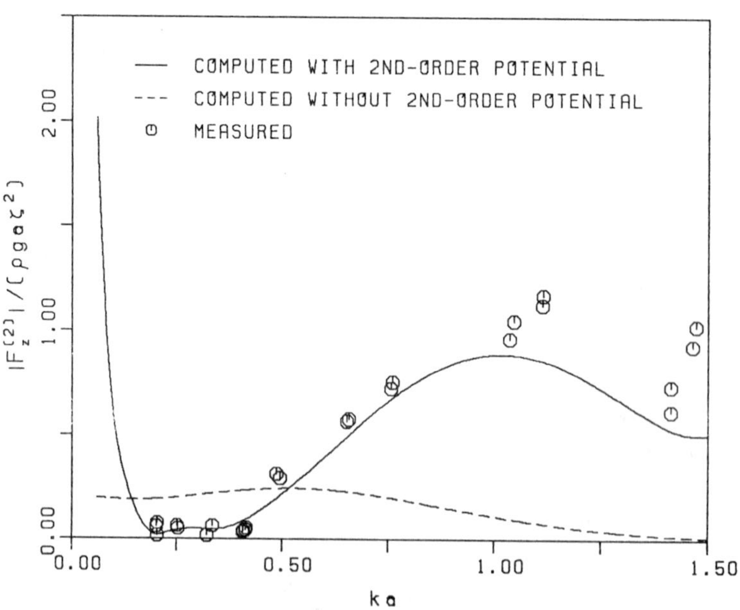


Figure 6: Comparison between computed and measured second-order double-frequency vertical forces on fixed truncated vertical cylinder.

An approximate second-order theory neglecting these effects is found to substantially underestimate the heave springing force. Agreement between theory and experiment is satisfactory except in the high-frequency range, where the forces are expected to be significantly influenced by the reflections of standing waves from the side walls of the tank.

# 3 Sum-frequency oscillations of a tension-leg platform

## 3.1 Equations of motion

Herein the motions of a TLP in regular waves are considered. In regular waves of frequency $\omega$, the motions of the TLP may be described in the form

$$\Xi_k = Re\left[\xi_k^{(1)}\mathrm{e}^{-i\omega t} + \xi_k^{(2)}\mathrm{e}^{-i2\omega t}\right] + \bar{\Xi}_k, \qquad k = 1, 2, \cdots, 6 \tag{69}$$

The first term in eqn (69) represents the wave-frequency responses due to the first-order wave exciting forces. The second term represents the sum-frequency responses due to the second-order wave exciting forces at twice the wave frequency. The last term represents the mean responses due to the wave drift and will not be considered here.

The equations of motion of the TLP may be derived by adding the effects of the tether system to the well-known equations of motion for a freely-floating structure. Omitting the detail of derivation, this leads to the equations of motion

$$\sum_{j=1}^{6}\left[-\omega^2(m_{kj} + M_{kj}^{(1)}) - i\omega N_{kj}^{(1)} - \frac{8}{3\pi}i\omega^2\hat{N}_{kj}^{(1)}\left|\xi_j^{(1)}\right| \right.$$
$$\left. + K_{kj} + L_{kj}\right]\xi_j^{(1)} = f_{kW}^{(1)}, \qquad k = 1, 2, \cdots, 6 \tag{70}$$

for the first-order wave-frequency motions, and

$$\sum_{j=1}^{6}\left[-4\omega^2(m_{kj} + M_{kj}^{(2)}) - i2\omega N_{kj}^{(2)} - \frac{8}{3\pi}i4\omega^2\hat{N}_{kj}^{(2)}\left|\xi_j^{(2)}\right| \right.$$
$$\left. + K_{kj} + L_{kj}\right]\xi_j^{(2)} = f_{kW}^{(2)} + g_{kW}^{(2)}, \qquad k = 1, 2, \cdots, 6 \tag{71}$$

for the second-order sum-frequency motions. In eqns (70) and (71), $m_{kj}$ are the inertia coefficients, $M_{kj}^{(1)}$, $M_{kj}^{(2)}$ are the (frequency-dependent) added inertia coefficients, $N_{kj}^{(1)}$, $N_{kj}^{(2)}$ are the linear (frequency-dependent) added damping coefficients, $\hat{N}_{kj}^{(1)}$, $\hat{N}_{kj}^{(2)}$ are the nonlinear viscous damping coefficients, $K_{kj}$ are the hydrostatic stiffness coefficients, $L_{kj}$ are the mooring stiffness coefficients, $f_{kW}^{(1)}$, $f_{kW}^{(2)}$ are the first- and second-order wave exciting forces, and $g_{kW}^{(2)}$ are the second-order forcing functions exerted by the tether system.

The frequency-dependent added inertia and damping coefficients, and the first-order wave exciting forces can be obtained from conventional first-order diffraction-radiation computations. The method for calculating the second-order wave exciting forces has been elaborated in the previous section. In the remainder of this section, discussion is limited to the evaluation of the terms in eqns (70) and (71) depending on the tether system.

### 3.2 Effects of tether system

Consider a floating body anchored to the seabed by $N$ vertical tethers of the lengths $l_m$, the extensional rigidities $k_m$ and the initial tensions $T_{0m}$ ($m = 1, 2, \cdots, N$), as shown in Fig. 7. It is assumed that the tethers are massless and that the effects of hydrodynamic forces acting on the tethers are negligible.

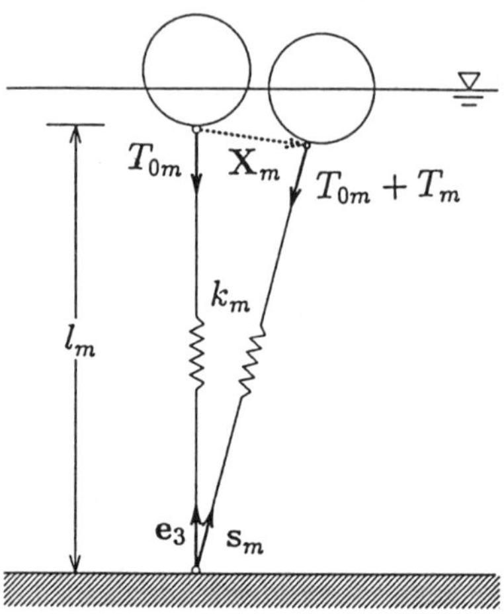


Figure 7: Sketch of tether system.

The mooring points on the body are denoted by the position vectors $\mathbf{x}_m$ relative to the body-fixed co-ordinate axes with an origin at the centre of gravity of the body. The displacement vector $\mathbf{X}_m$ of the point $\mathbf{x}_m$, referred to the space-fixed co-ordinate axes, may then be expressed as

$$\mathbf{X}_m = X_{1m}\mathbf{e}_1 + X_{2m}\mathbf{e}_2 + X_{3m}\mathbf{e}_3 = \mathbf{U} + (\underline{D} - \underline{I})\mathbf{x}_m \tag{72}$$

which may conveniently be expanded into

$$\mathbf{X}_m = \varepsilon\mathbf{X}_m^{(1)} + \varepsilon^2\mathbf{X}_m^{(2)} + O(\varepsilon^3) \tag{73}$$

The first- and second-order motion vectors $\mathbf{X}_m^{(1)}$, $\mathbf{X}_m^{(2)}$ of the point $\mathbf{x}_m$ may then be written as

$$\mathbf{X}_m^{(1)} = X_{1m}^{(1)}\mathbf{e}_1 + X_{2m}^{(1)}\mathbf{e}_2 + X_{3m}^{(1)}\mathbf{e}_3 = \mathbf{U}^{(1)} + \mathbf{W}^{(1)} \times \mathbf{x}_m \tag{74}$$

$$\mathbf{X}_m^{(2)} = X_{1m}^{(2)}\mathbf{e}_1 + X_{2m}^{(2)}\mathbf{e}_2 + X_{3m}^{(2)}\mathbf{e}_3 = \mathbf{U}^{(2)} + \mathbf{W}^{(2)} \times \mathbf{x}_m + \underline{H}^{(2)}\mathbf{x}_m \tag{75}$$

Equations (74) and (75) give the relations between the displacements of the mooring points and the motions of the body.

The extension $e_m$ and the resulting fluctuation of tether force $T_m$ may be obtained from

$$e_m = |l_m \mathbf{e}_3 + \mathbf{X}_m| - l_m = \sqrt{X_{1m}^2 + X_{2m}^2 + (l_m + X_{3m})^2} - l_m \tag{76}$$

$$T_m = k_m e_m \tag{77}$$

Introducing eqns (73)–(75) into eqn (76), and collecting terms of the lowest powers of $\varepsilon$, it follows that

$$e_m = \varepsilon e_m^{(1)} + \varepsilon^2 e_m^{(2)} + O(\varepsilon^3) \tag{78}$$

where

$$e_m^{(1)} = X_{3m}^{(1)}, \quad e_m^{(2)} = X_{3m}^{(2)} + \frac{1}{2l_m}\left(X_{1m}^{(1)2} + X_{2m}^{(1)2}\right) \tag{79}$$

Hence

$$T_m = \varepsilon T_m^{(1)} + \varepsilon^2 T_m^{(2)} + O(\varepsilon^3) \tag{80}$$

where

$$T_m^{(1)} = k_m X_{3m}^{(1)}, \quad T_m^{(2)} = k_m \left[X_{3m}^{(2)} + \frac{1}{2l_m}\left(X_{1m}^{(1)2} + X_{2m}^{(1)2}\right)\right] \tag{81}$$

Equation (81) implies that not only the second-order vertical motions but also the quadratic products of the first-order horizontal motions contribute to the second-order tether forces.

The resultant tether forces $\mathbf{G}$ exerted on the body may be obtained from

$$\mathbf{G} = -\sum_{m=1}^{N} (T_{0m} + T_m)\,\mathbf{s}_m \tag{82}$$

where $\mathbf{s}_m$ is the unit vector in the direciton of the tether in the displaced position

$$\mathbf{s}_m = \frac{1}{l_m + e_m}(l_m \mathbf{e}_3 + \mathbf{X}_m) \tag{83}$$

Introducing eqns (73)–(75) and (78), (79) into eqn (83), it follows that

$$\mathbf{s}_m = \mathbf{e}_3 + \varepsilon \mathbf{s}_m^{(1)} + \varepsilon^2 \mathbf{s}_m^{(2)} + O(\varepsilon^3) \tag{84}$$

where

$$\mathbf{s}_m^{(1)} = \frac{X_{1m}^{(1)}}{l_m}\mathbf{e}_1 + \frac{X_{2m}^{(1)}}{l_m}\mathbf{e}_2 \tag{85}$$

$$\mathbf{s}_m^{(2)} = \left(\frac{X_{1m}^{(2)}}{l_m} - \frac{X_{1m}^{(1)}X_{3m}^{(1)}}{l_m^2}\right)\mathbf{e}_1 + \left(\frac{X_{2m}^{(2)}}{l_m} - \frac{X_{2m}^{(1)}X_{3m}^{(1)}}{l_m^2}\right)\mathbf{e}_2 - \frac{X_{1m}^{(1)2} + X_{2m}^{(1)2}}{2l_m^2}\mathbf{e}_3 \tag{86}$$

Substituting eqns (80), (81) and (84)–(86) into eqn (82), and collecting powers of $\varepsilon$, the expressions for **G** may be derived in the form

$$\mathbf{G} = \mathbf{G}^{(0)} + \varepsilon \mathbf{G}^{(1)} + \varepsilon^2 \mathbf{G}^{(2)} + O(\varepsilon^3) \tag{87}$$

where

$$\mathbf{G}^{(0)} = -\sum_{m=1}^{N} T_{0m} \mathbf{e}_3 \tag{88}$$

$$\mathbf{G}^{(1)} = -\sum_{m=1}^{N} \left( \frac{T_{0m}}{l_m} X_{1m}^{(1)} \mathbf{e}_1 + \frac{T_{0m}}{l_m} X_{2m}^{(1)} \mathbf{e}_2 + k_m X_{3m}^{(1)} \mathbf{e}_3 \right) \tag{89}$$

$$\begin{aligned} \mathbf{G}^{(2)} = & -\sum_{m=1}^{N} \left( \frac{T_{0m}}{l_m} X_{1m}^{(2)} \mathbf{e}_1 + \frac{T_{0m}}{l_m} X_{2m}^{(2)} \mathbf{e}_2 + k_m X_{3m}^{(2)} \mathbf{e}_3 \right) \\ & -\sum_{m=1}^{N} \frac{1}{l_m} \left( k_m - \frac{T_{0m}}{l_m} \right) \Big[ X_{1m}^{(1)} X_{3m}^{(1)} \mathbf{e}_1 + X_{2m}^{(1)} X_{3m}^{(1)} \mathbf{e}_2 \\ & + \frac{1}{2} \left( X_{1m}^{(1)2} + X_{2m}^{(1)2} \right) \mathbf{e}_3 \Big] \end{aligned} \tag{90}$$

A similar approach may be used to derive the corresponding expressions for the resulting moments **Q** of the tether forces on the body. These moments are obtained from

$$\mathbf{Q} = -\sum_{m=1}^{N} (\underline{D}\mathbf{x}_m) \times [(T_{0m} + T_m)\, \mathbf{s}_m] \tag{91}$$

In eqn (91), $\underline{D}\mathbf{x}_m$ represents the displaced position vector of the mooring point $\mathbf{x}_m$, referred to the co-ordinate axes parallel to the space-fixed co-ordinate axes but with an origin at the centre of gravity. Using eqns (72)–(75), $\underline{D}\mathbf{x}_m$ may be written as

$$\underline{D}\mathbf{x}_m = \mathbf{x}_m + \varepsilon \left( \mathbf{W}^{(1)} \times \mathbf{x}_m \right) + \varepsilon^2 \left( \mathbf{W}^{(2)} \times \mathbf{x}_m + \underline{H}^{(2)} \mathbf{x}_m \right) + O(\varepsilon^3) \tag{92}$$

Substituting eqns (80), (81), (84)–(86) and (92) into eqn (91), the final expressions for **Q** may be derived in the form

$$\mathbf{Q} = \mathbf{Q}^{(0)} + \varepsilon \mathbf{Q}^{(1)} + \varepsilon^2 \mathbf{Q}^{(2)} + O(\varepsilon^3) \tag{93}$$

where

$$\mathbf{Q}^{(0)} = -\sum_{m=1}^{N} \mathbf{x}_m \times T_{0m} \mathbf{e}_3 \tag{94}$$

$$\begin{aligned} \mathbf{Q}^{(1)} = & -\sum_{m=1}^{N} \Big[ \mathbf{x}_m \times \left( \frac{T_{0m}}{l_m} X_{1m}^{(1)} \mathbf{e}_1 + \frac{T_{0m}}{l_m} X_{2m}^{(1)} \mathbf{e}_2 + k_m X_{3m}^{(1)} \mathbf{e}_3 \right) \\ & + \mathbf{W}^{(1)} \times \mathbf{x}_m \times T_{0m} \mathbf{e}_3 \Big] \end{aligned} \tag{95}$$

$$\mathbf{Q}^{(2)} = -\sum_{m=1}^{N} \Big[ \mathbf{x}_m \times \left( \frac{T_{0m}}{l_m} X_{1m}^{(2)} \mathbf{e}_1 + \frac{T_{0m}}{l_m} X_{2m}^{(2)} \mathbf{e}_2 + k_m X_{3m}^{(2)} \mathbf{e}_3 \right)$$

$$+\mathbf{W}^{(2)} \times \mathbf{x}_m \times T_{0m}\mathbf{e}_3\Big] - \sum_{m=1}^{N}\left\{\frac{1}{l_m}\left(k_m - \frac{T_{0m}}{l_m}\right)\mathbf{x}_m \times \left[X_{1m}^{(1)}X_{3m}^{(1)}\mathbf{e}_1\right.\right.$$

$$\left.+X_{2m}^{(1)}X_{3m}^{(1)}\mathbf{e}_2 + \frac{1}{2}\left(X_{1m}^{(1)2} + X_{2m}^{(1)2}\right)\mathbf{e}_3\right] + \mathbf{W}^{(1)} \times \mathbf{x}_m \times \left(\frac{T_{0m}}{l_m}X_{1m}^{(1)}\mathbf{e}_1\right.$$

$$\left.\left.+\frac{T_{0m}}{l_m}X_{2m}^{(1)}\mathbf{e}_2 + k_m X_{3m}^{(1)}\mathbf{e}_3\right) + \underline{H}^{(2)}\mathbf{x}_m \times T_{0m}\mathbf{e}_3\right\} \tag{96}$$

The expressions for the resultant tether forces and moments obtained above may be combined more conveniently in the form

$$G_k = G_k^{(0)} + \varepsilon G_k^{(1)} + \varepsilon^2 G_k^{(2)} + O(\varepsilon^3), \qquad k = 1, 2, \cdots, 6 \tag{97}$$

$G_k^{(0)}$ represents the resultant forces and moments due to the initial tensions, which are balanced with the surplus buoyancy. Using eqns (74), (75) together with eqns (7)–(9), $G_k^{(1)}$ and $G_k^{(2)}$ may be expressed in terms of the motions of the body as

$$G_k^{(1)} = -L_{kj}\Xi_j^{(1)}, \qquad G_k^{(2)} = -L_{kj}\Xi_j^{(2)} + G_{kW}^{(2)} \tag{98}$$

where $G_{kW}^{(2)}$ represents the terms consisting of the quadratic products of the first-order body motions. $-L_{kj}\Xi_j^{(n)}$ in eqn (98) correspond to the static restoring forces due to the tether system. $G_{kW}^{(2)}$ provides the additional forcing functions exciting the second-order motions. The equations of motion of a TLP may then be obtained by adding these effects to the well-known equations of motion for a freely-floating structure, as given in eqns (70) and (71).

## 4. Experiment

### 4.1 Test set-up

In order to validate the theoretical approach described in the previous sections, model tests have been performed in regular waves with a small scale model of four-column TLP. The principal particulars of the model are presented in Fig. 8 and Table 2. The model consists of four circular columns and four submerged rectangular pontoons, both made of acrylite, and is anchored to the tank bottom by four vertical tethers, composed of a combination of stainless wires and steel springs and connected to the centres of column bottoms. At the bottom end of each tether, a ring-shaped strain-gauge force transducer was attached to measure the tension in the tether. On the top of the deck, a white wooden pole was installed on which two black points were marked as the targets to trace the movement of the model by means of a contactless image tracking system.

The tests were made in the wave tank at Nagoya University. This facility has a width of 2m, a maximum water depth of 1.5m and a distance of approximately 15m between the wavemaker and the beach. The model was set up about half way along the length of the tank. The depth of the tank is not large enough to scale the water depth and the tether lengths of the prototype structure. The axial stiffnesses of the tethers were therefore adjusted such that the resonant vertical-plane oscillations due to the springing forces are well excited within the frequency range at which the wavemaker could produce stable waves.

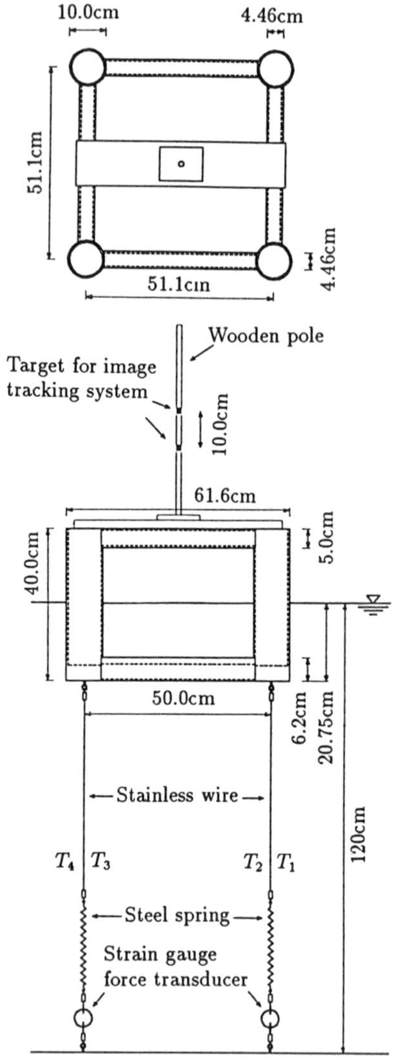


Figure 8: TLP model and test set-up.

## 4.2 Free oscillation tests in still water

Prior to the tests in waves, free oscillation tests were performed in still water to measure the natural frequencies and dampings for the surge, heave and pitch modes. The linear and nonlinear damping coefficients were determined from the differences between two subsequent amplitudes of oscillation, as described in the following.

Table 2: Principal particulars of TLP model

| Description | Value | Unit |
|---|---|---|
| Spacing between column centre lines | 51.1 | cm |
| Column diameter | 10 | cm |
| Operating draft | 20.75 | cm |
| Submergence depth of COG | 4.217 | cm |
| Total mass | 9.495 | kg |
| Pitch moment of inertia | 7221.7 | kg cm$^2$ |
| Displacement | 11.351 | kg |
| Number of mooring tethers | 4 | – |
| Length of the tether | 99.25 | cm |
| Axial stiffness of a tether | 1011 | gf/cm |
| Pre-tension in a tether | 464 | gf |
| Water depth | 120 | cm |

The equation of motion for a single degree-of-freedom system undergoing free oscillation can be expressed as

$$(m + M)\ddot{y} + N\dot{y} + \hat{N}\,|\dot{y}|\,\dot{y} + (K + L)y = 0 \tag{99}$$

where $m$ is the inertia coefficient, $M$ is the added inertia coefficient, $N$ is the linear added damping coefficient, $\hat{N}$ is the nonlinear viscous damping coefficient, $K$ is the hydrostatic stiffness coefficient and $L$ is the mooring stiffness coefficient.

Let the subsequent amplitudes of oscillation be denoted by $y_1, y_2, \cdots, y_{n-1}, y_n, y_{n+1}, \cdots$. The ratio of two subsequent amplitudes can then be expressed as (Hooft, 1982)

$$\frac{y_{n+2} + y_{n+1}}{y_{n+1} + y_n} = \exp\left[-\frac{\pi N}{2(m+M)}\right]\left[1 - \frac{2\hat{N}}{3(m+M)}(y_n + y_{n+1})\right] \tag{100}$$

Equation (100) may be rewritten in the form

$$z_{n+1} - z_n = -\left\{1 - \exp\left[-\frac{\pi N}{2(m+M)}\right]\right\} z_n - \frac{2}{3}\exp\left[-\frac{\pi N}{2(m+M)}\right]\hat{N} z_n^2 \tag{101}$$

where $z_n = y_n + y_{n+1}$. With the known inertia and added inertia coefficients, $m$ and $M$, the linear and nonlinear damping coefficients, $N$ and $\hat{N}$, can then be determined from the curve $z_{n+1} - z_n$ versus $z_n$, which is best fitted by means of the least square method.

### 4.3 Tests in regular waves

The tests were carried out in regular waves for ranges of wave frequencies (0.5–1.5Hz) and amplitudes (1.5–5cm). The following quantities were measured:

- Wave excitation forces on a fixed TLP model by means of load cells.
- Motion responses of a moored TLP model by means of a contactless image tracking system.
- Tether forces of a moored TLP model by means of ring-shaped strain-gauge force transducers.
- Incident wave height by means of a servo-type wave prove positioned 2.5m in front of the model.

The signals recorded by a cassette-type data recording system were converted to digital data by means of a A-D converter. A harmonic analysis based on the Fast Fourier Transform algorithm was peformed to filter out the first and second harmonic components of waves, forces and responses.

# 5 Discussion of results

## 5.1 Natural frequencies and dampings

The surge, heave and pitch natural frequencies and the corresponding damping coefficients measured from free oscillation tests in still water are presented in Table 3. These damping values were determined from the differences between two subsequent amplitudes of oscillation, which were best fitted by means of least square method, as shown in Fig. 9.

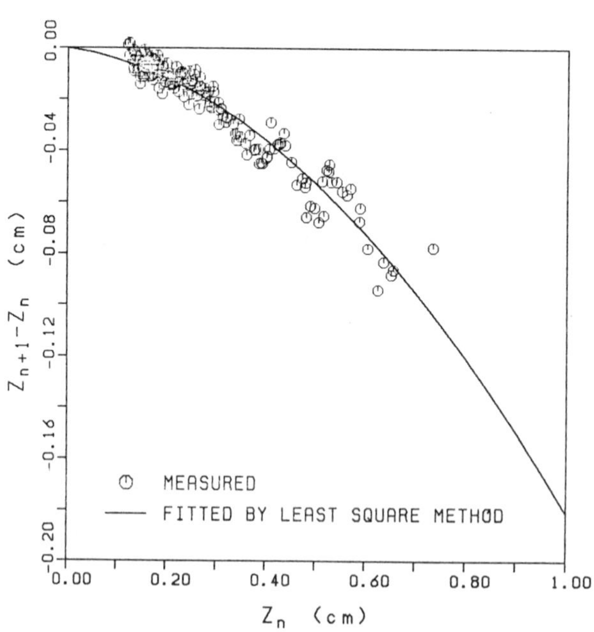


Figure 9: Evaluation of heave damping coefficients for TLP model by free oscillation decay test.

Table 3: Measured natural frequencies and damping coefficients for TLP model

| Mode | Natural frequency | Damping coefficient | |
|---|---|---|---|
| | | Linear | Nonlinear |
| Surge | 0.175 Hz | 0.608 gf s/cm | 0.813 gf $s^2/cm^2$ |
| Heave | 2.49 Hz | 5.25 gf s/cm | 4.22 gf $s^2/cm^2$ |
| Pitch | 2.39 Hz | 963 gf s | $1.54\times10^5$ gf $s^2/cm$ |

### 5.2 Wave excitation forces on fixed TLP model

The measured wave exciting forces on the fixed TLP model were compared with the computations based on the second-order wave diffraction theory. The element idealisation shown in Fig. 3 was employed for the calculation based on the hybrid integral-equation method. 264, 720 and 16 quadratic isoparametric elements were used to model the underwater hull surface, the free surface and the seabed, respectively. The eigenfunction expansion for the outer fluid domain was truncated with 7 Fourier harmonics and 14 evanescent modes. Two wave directions, i.e. the head wave ($\alpha = 0$) and the bow-quartering wave ($\alpha = \pi/4$), were considered. Figures 10 and 11 show the comparison for the second harmonic components of heave wave force. The solid and dotted lines in these figures indicate the results computed with and without the second-order potential effects. It is observed that the contribution of the second-order potential dominates the total second-order wave forces. The approximate second-order theory neglecting these effects is found to substantially underestimate the heave springing forces. The correlation between theory and experiment is considered to be generally satisfactory, in view of the small magnitudes of the forces being measured. Large discrepancies observed in the high frequency range are thought to be caused by the reflections of standing waves from the side walls of the tank.

### 5.3 Motion responses and tether forces of moored TLP model

The measured motion responses and tether forces of the moored TLP model were also compared with the theoretical predictions based on the measured damping values. Figures 12 and 13 show the comparison for the second harmonic components of heave response and tether force in head waves. It is observed that the response curves for the heave motion and tether force exhibit sharp peaks at the frequencies of 1.25Hz and 1.2Hz. The fact that these frequencies coincide with half the heave and pitch natural frequencies suggests the occurrence of the resonant oscillations excited by the second-order wave loads at twice the wave frequency. Except in the vicinity of the resonant frequencies, the computed results are found to correlate well with the experimental data. However, the large discrepancies observed near the resonant frequencies suggest the need for improved estimates of the damping forces (including mechanical and hydrodynamic).

The contribution of the additional second-order forcing functions due to the tether system (set-down effects) is also shown in Figs 12 and 13. It is observed that this

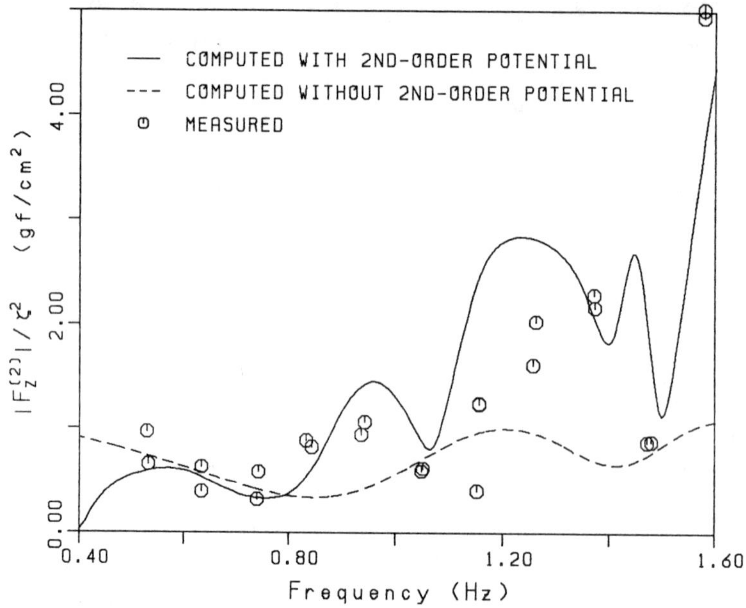


Figure 10: Comparison between computed and measured second harmonic components of heave wave force on fixed TLP in head waves.

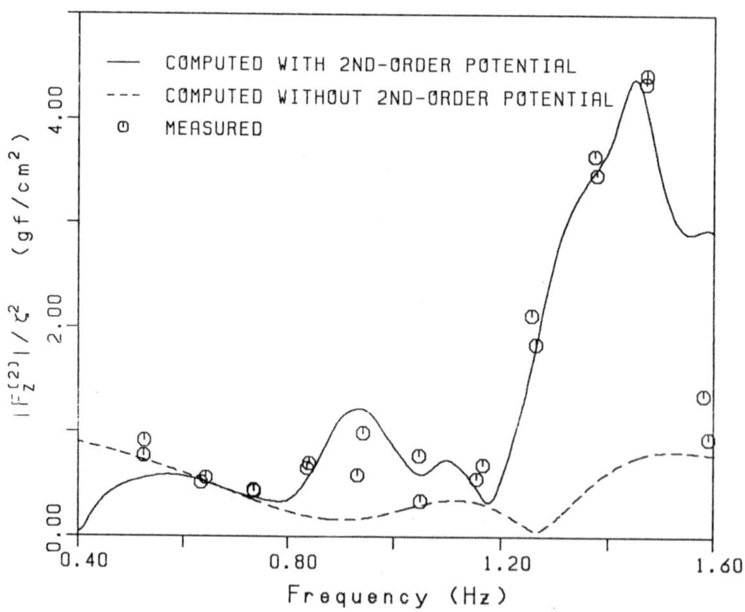


Figure 11: Comparison between computed and measured second harmonic components of heave wave force on fixed TLP in bow-quartering waves.

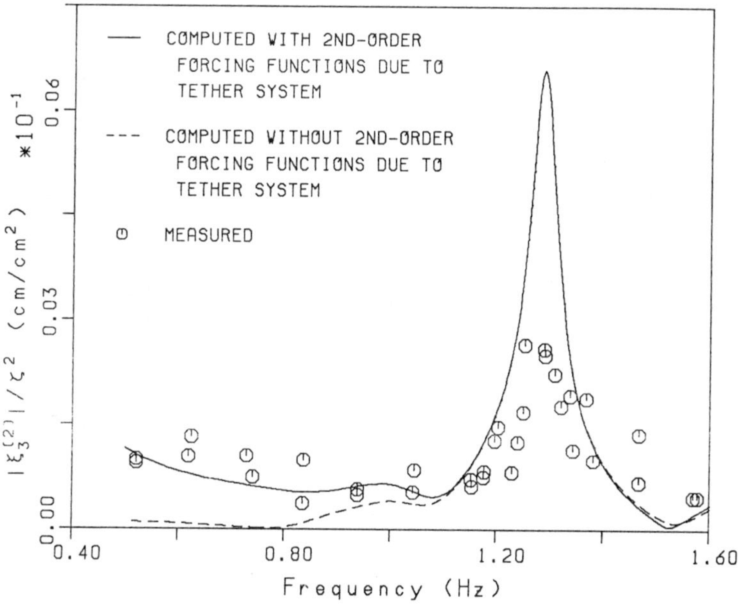


Figure 12: Comparison between computed and measured second harmonic components of heave response of moored TLP in head waves.

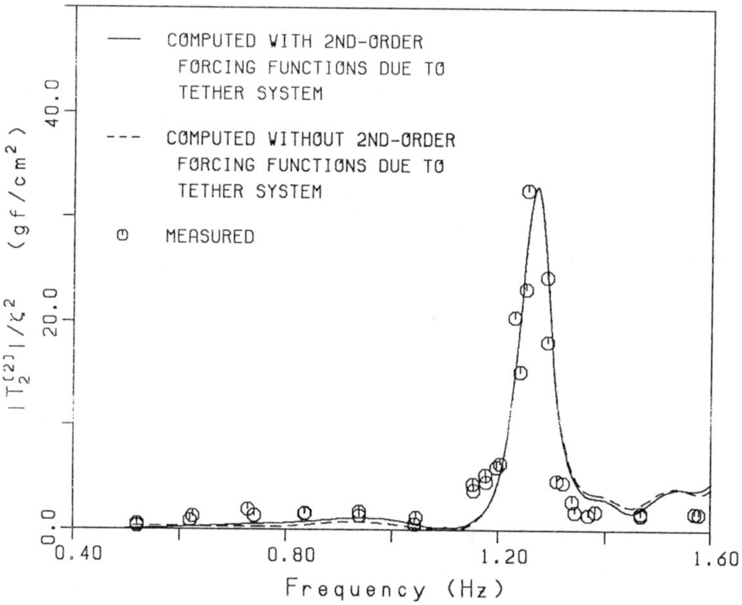


Figure 13: Comparison between computed and measured second harmonic components of fore tether force $T_2$ of moored TLP in head waves.

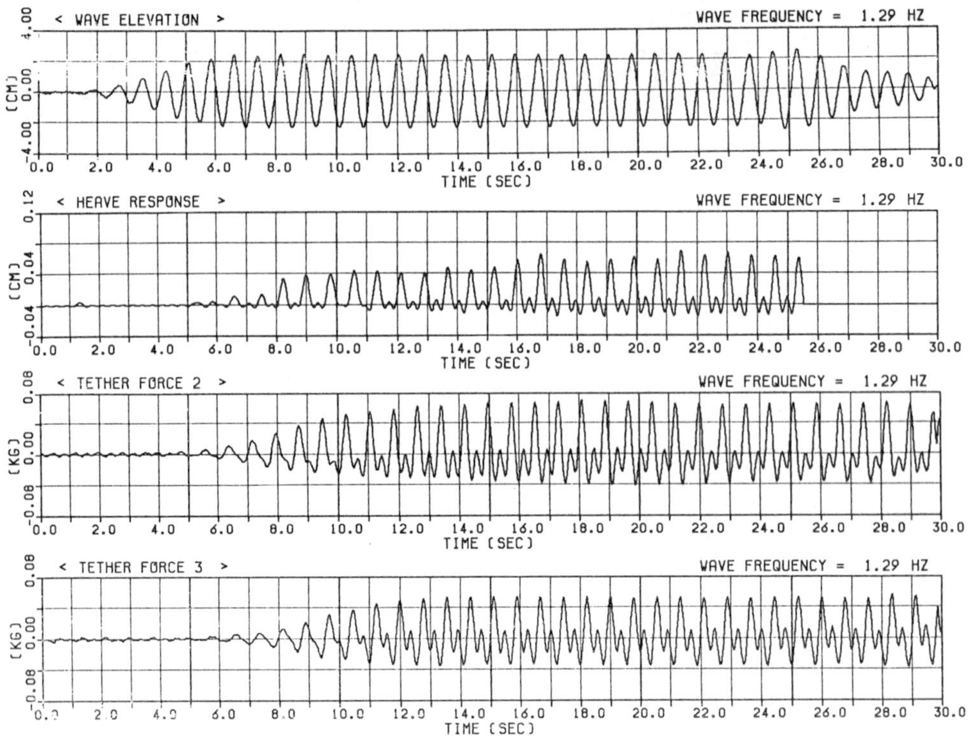


Figure 14: Measured time histories of heave response and tether forces of moored TLP in head waves (frequency= 1.29Hz, wave amplitude= 2.35cm).

contribution is quite significant on the heave response at low frequencies. This is due to large horizontal vessel motions at these frequencies and relatively shallow water depth considered here. However, the effect on the tether force is found to be small.

Figure 14 illustrates an example of measured time histories of heave response and tether tensions near the resonant frequency. The existence of significant second harmonic components are clearly observed, which are comparable in magnitude with the first harmonic wave-frequency components.

## 6 Conclusions

A complete second-order diffraction theory has been developed for computing the second-order sum-frequency wave loads on three-dimensional floating bodies in regular waves. The theory was applied to predict the springing oscillation of a TLP. In order to validate the theoretical predictions, model tests have been performed with a small scale model of four-column TLP. Satisfactory agreement was observed between analysis and experimental data, thus confirming the validity of the theory. The conclusions arising from the present study are summarised as follow:

- The vertical-plane motions and resulting tether forces of a TLP include the second-order sum-frequency resonant components, which are comparable in magnitude with the first-order wave-frequency responses.

- The nonlinear sum-frequency wave exciting forces, often called 'springing' forces, provide an important source exciting such resonant oscillations. These excitation forces can be predicted easily based on the second-order wave diffraction theory.

- The contribution of the second-order potential dominates the total second-order wave forces. The approximate second-order theory using second-order contributions due to the first-order potentials only substantially underestimates the springing wave loads.

- The theoretical approach described in this chapter provides an effective means for predicting the vertical-plane springing oscillation and tether forces of a TLP, although further research is needed for better estimates of the damping forces.

The theory developed in this chapter may readily be extended to calculate the sum- and difference-frequency wave loads and responses in irregular waves (Matsui, 1989).

## References

Abramowitz, M. & Stegun, I.A. (1972) *Handbook of Mathematical Functions*, 9th ed., Dover, New York.

Chau, F.P. (1989) The second-order velocity potential for diffraction of waves by fixed offshore structures, Ph. D. Thesis, University College London.

De Boom, W.C., Pinkster, J.A. & Tan, P.S.G. (1984) Motion and tether force prediction for a TLP, *J. Waterways, Port, Coastal and Ocean Eng., Proc. ASCE*, **110**(4), 472-486.

Eatock Taylor, R. & Hung, S.M. (1987) Second order diffraction forces on a vertical cylinder in regular waves, *Appl. Ocean Res.*, **9**(1), 19-30.

Hooft, J.P., (1982) *Advanced Dynamics of Marine Structures*, John Wiley & Sons, New York.

Kim, M.H. (1991) Second-order sum-frequency wave loads on large-volume structures, *Appl. Ocean Res.*, **13**(6), 287-296.

Kim, M.H. & Yue, D.K.P. (1988) The nonlinear sum-frequency wave excitation and response of a tension-leg platform, *Proc. 5th Int. Conf. on the Behaviour of Offshore Structures (BOSS'88)*, Trondheim, Norway, pp.687-704.

Kim, M.H. & Yue, D.K.P. (1989) The complete second-order diffraction solution for an axisymmetric body. Part 1. Monochromatic incident waves, *J. Fluid Mech.*, **200**, 235-264.

Kim, M.H. & Yue, D.K.P. (1990) The complete second-order diffraction solution for an axisymmetric body. Part 2. Bichromatic incident waves and body motions, *J. Fluid Mech.*, **211**, 557-593.

Lighthill, M.J. (1979) Waves and hydrodynamic loading, *Proc. 2nd Int. Conf. on the Behaviour of Offshore Structures (BOSS'79)*, London, Vol.1, pp.1-40.

Matsui, T. (1989) Computation of slowly varying second-order hydrodynamic forces on floating structures in irregular waves, *J. Offshore Mech. and Arctic Eng., Trans. ASME*, **111**, 223-232.

Matsui, T. & Kato, K. (1991) The analysis of wave-induced dynamics of ocean platforms by hybrid integral-equation method, *Int. J. Offshore and Polar Eng., Trans. ISOPE*, **1**(2), 146-153.

Matsui, T., Sakoh, Y. & Nozu, T. (1992) Experimental measurement of second-order sum-frequency oscillations of a tension-leg platform, *Proc. 6th Int. Conf. on the Behaviour of Offshore Structures (BOSS'92)*, London, Vol.1, pp.584-594.

Matsui, T., Sakoh, Y. & Nozu, T. (1993) Second-order sum-frequency oscillations of tension-leg platforms: prediction and measurement, *Appl. Ocean Res.*, **15**(2), 107-118.

Matsui, T., Suzuki, T. & Sakoh, Y. (1992) Second-order diffraction forces on floating three-dimensional bodies in regular waves, *Int. J. Offshore and Polar Eng., Trans. ISOPE*, **2**(3), 175-185.

Molin, B. (1979) Second order diffraction loads upon three-dimensional bodies, *Appl. Ocean Res.*, **1**(4), 197-202.

Molin, B. & Chen, X.B. (1990) Calculation of second-order sum-frequency loads on TLP hulls, Interim Report on IFP's Part of the Project: Vertical Resonant Motions of Tension Leg Platforms, FNS Project No.24841.

Ogilvie, T.F. (1983) Second-order hydrodynamic effects on ocean platforms, *Int. Workshop on Ship and Platform Motions*, Berkeley, California, pp.205-265.

Ogilvie, T.F. & Tuck, E.O. (1969) A rational strip theory for ship motions : Part 1, Rept. No. 013, Dept. of Naval Architecture and Marine Eng., Univ. of Michigan.

Petrauskas, C. & Liu, S.V. (1987) Springing force response of a tension leg platform, *Proc. 19th Offshore Technology Conf.*, Paper No. OTC 5458, Houston, Texas, Vol.2, pp.333-341.

Rahman, M. (1984) Wave diffraction by large offshore structures: an exact second order theory, *Appl. Ocean Res.*, **6**(2), 90-100.

Yoshida, K., Yoneya, T., Oka, N. & Ozaki, M. (1981) Motions and leg tensions of tension leg platforms, *Proc. 13th Offshore Technology Conf.*, Paper No. OTC 4073, Houston, Texas, Vol.3, pp.75-87.

# Chapter 3

# Some radiation and diffraction problems in the linearised theory of water waves

**B.N. Mandal,**[a] **S. Banerjea**[b]

[a] *Physics and Applied Mathematics Unit, Indian Statistical Institute, 203 B.T. Road, Calcutta 700035, India*

[b] *Department of Mathematics, Jadavpur University, Calcutta 700032, India*

**Abstract**

Assuming linearised theory of water waves a class of two-dimensional radiation and diffraction problems have been studied here. This class of problems reduce to boundary value problems (BVP) associated with Laplace's equation for the determination of velocity potential. The BVPs satisfy a Churchill type boundary condition at one boundary (the free surface) and Neumann type condition at other boundaries. A simplified perturbation technique is employed directly to the governing equations to study the BVPs involving nearly vertical plate, submerged nearly circular cylinder. Some special BVPs arising here are also considered. The mathematical techniques used here involve integral equation approach involving suitable applications of Green's integral theorem or Havelock's expansion of water wave potential.

# 1 Introduction

The phenomena of wave propagation are encountered in almost all branches of mathematical physics. It is somewhat difficult to give a precise definition of what constitutes a wave, but to cover the whole range of wave phenomenon without being restrictive, it can be stated intuitively that a wave is any recognizable signal that is transferred from one part of a medium to another with a recognizable velocity of propagation. The signal may be any feature of disturbance such as maximum or abrupt change in some physical quantities provided it is clearly recognizable and its location at any time can be determined. This perhaps constitutes a general description of wave phenomenon although different features are important in different types of waves, (cf. Whitham (1974) p2). According to Sommerfeld, 'ever since waves were studied, water waves have served the natural scientist as a model for wave theory in general (cf. Segel (1977) p 299)'. Consequently the wave phenomena in water with a free surface under the action of gravity has attracted the attention of a number of mathematicians and physicists such as Cauchy (1759-1857), Poisson (1781-1840), Lagrange (1736-1813), Airy (1801-1892), Stokes (1819-1903), Kelvin (1824-1907) and many others. This has resulted in a systematic development of the theory of water waves from the latter half of the eighteenth century. This theory has provided a background for a somewhat rich development of mathematical concepts and techniques and consequently it now constitutes an important branch of applied mathematics as well as mathematical physics. It is most varied and fascinating, and includes a wide range of natural phenomena in oceans, rivers and lakes.

The theory of water waves is mostly concerned with elucidation of some general aspects of wave motion as well as the prediction of the behaviour of waves in the presence of some special configurations of interest to oceanographers, hydraulic engineers and ship designers. Unfortunately, even the simplest problems appear to be difficult to tackle mathematically and thus it is necessary to make some simplified assumptions about the medium as well as the wave motion. The nature of assumptions made in treating a particular problem provides a natural way to classify the theory of water waves. Primarily, some assumptions are made concerning the physical properties of the fluid. Consequently, water is assumed to be an incompressible, inviscid and homogeneous fluid. So if one neglects viscosity and assumes irrotational motion, the problems concerning gravity waves reduce to finding solutions to Laplace's equation in the fluid region with appropriate boundary conditions, which again are to be simplified by making some other assumptions. For example, the boundary condition at the free surface is in general nonlinear and pose mathematical difficulties and consequently the problems become unmanageable. Thus some other assumptions not concerning the nature of the fluid but concerning the nature of fluid motion and its generation have to be made which will obviously reduce the mathematical complicity without restricting the physical significance. As a consequence, two approximate theories have been developed from the basic hydrodynamic theory with the nomenclatures linearised theory and shallow water theory. We will consider here only the linearised theory. This theory is based on the assumption that the amplitude of wave motion is small compared to the wave length. It is developed from the basic general theory of fluid mechanics by assuming velocity components and free surface elevation, together with their partial derivatives, to be of small quantities so that their products and powers can be neglected.

There are varied classes of problems which may be considered under the linearised theory. Amongst others, they include radiation and diffraction problems involving bodies of various geometrical shapes present in water. If a time-harmonic progressive wave train is incident on a body or bodies present in water, a part of it is reflected and the remaining part is transmitted. The study of the ensuing motion and the determination of the reflection and transmission coefficients is referred to as the diffraction or scattering problem. Again, if one or more bodies present in water undergo some sort of small oscillations, then waves are generated in the water region and radiate away from the body or bodies. The study of the ensuing motion and the determination of wave amplitudes far away from the body on both its sides is referred to as the radiation problem. These two classes of problems have considerable practical applications in the ship building industry, construction of harbours, off-shore structures, etc. It may be noted here that closed form solutions to the problems belonging to these classes are available in the literature only for a limited number of cases, such as when the body is in the form of a thin vertical plate, a thin plane barrier inclined at a special angle for the water wave diffraction problem and also for two-dimensional motion in deep water. However, for problems involving bodies of other geometrical shapes, approximate solutions can be obtained by using an integral equation procedure based on Green's integral theorem.

The present article is mainly concerned with some problems of water wave radiation and diffraction involving bodies of some special configurations, namely nearly vertical thin barriers and nearly circular cylinders. The problems are studied by employing various appropriate mathematical techniques. A brief description of the general formulation of water wave radiation and diffraction problems is given below.

## 1.1 Radiation problem

If a body present in water is forced to execute a small prescribed motion, a wave motion is set up in the water region which radiates away from the body. If the motion is two-dimensional and time-harmonic with angular frequency $\sigma$ and is described by a velocity potential which is the real part of $\phi(x, y)\ e^{-i\sigma t}$, where the y-axis is taken vertically downward into the water and y=0 plane is the position of the free surface at rest, then $\phi$ satisfies

$$\nabla^2 \phi = 0 \quad \textit{in the fluid region,} \tag{1.1}$$

the free surface condition

$$K\,\phi + \phi_y = 0 \textit{ on the free surface} \tag{1.2}$$

where K $= \sigma^2/g$, g being the acceleration due to gravity, the condition on the body given by

$$\frac{\delta\phi}{\delta n} = \textit{prescribed on } W \tag{1.3}$$

where W is the wetted part of the body and n is the outward drawn normal to the body, the bottom condition

$$\nabla\phi \longrightarrow 0 \quad as\ y \longrightarrow \infty \tag{1.4}$$

and finally the radiation condition

$$\phi \rightarrow \begin{cases} A\, e^{-Ky+iKx} & as\ x \rightarrow \infty, \\ Be^{-Ky-iKx} & as\ x \rightarrow -\infty \end{cases} \tag{1.5}$$

where A, B denote the amplitudes of radiated waves at a large distance from the body on its two sides.

## 1.2 Diffraction problem

If a train of time-harmonic progressive waves represented by Re $\{\phi^{inc}(x,y)e^{-i\sigma t}\}$ is incident from negative infinity on a stationary body present in water, a part of it is reflected by the body and the remaining part is transmitted over or below it. If the motion is two-dimensional and described by a velocity potential Re $\{\phi(x,y)e^{-i\sigma t}\}$, then $\phi$ satisfies

$$\nabla^2\phi = 0 \ \textit{in the fluid region,} \tag{1.6}$$

the free surface condition

$$K\phi + \phi_y = 0 \ \textit{on the free surface,} \tag{1.7}$$

the condition on the body given by

$$\frac{\delta\phi}{\delta n} = 0 \textit{ on } W, \tag{1.8}$$

W being the wetted part of the body and n being outward drawn normal to it, the bottom condition

$$\nabla\phi \longrightarrow 0 \ as\ y \rightarrow \infty \tag{1.9}$$

and finally

$$\phi \rightarrow \begin{cases} \phi^{inc}(x,y) + R\phi^{inc}(-x,y) as\ x \rightarrow -\infty, \\ T\phi^{inc}(x,y) \qquad\qquad\qquad as\ x \rightarrow \infty \end{cases} \tag{1.10}$$

where R and T denote the reflection and transmission coefficients which are unknown, and the incident field is given by

$$\phi^{inc}(x,y) = e^{-Ky+iKx}.$$

If there is a submerged sharp edge of the body then for both diffraction and radiation problems, another additional condition on $\phi$ has to be supplemented near this edge. This is known as the edge condition. Following the arguments of Karp and Karal (1962), a brief description of this edge condition is given below.

## 1.3 Edge condition

Let $\phi$ (x,y) be a harmonic function in the x-y plane except on the barrier (y = 0, $0 \leq x < \infty$) having an edge at the origin, and on which $\phi$ satisfies the Neumann condition on its two sides. It is required to determine the behaviour of $\phi$ near the origin. Using the polar co-ordinates system $(r, \theta)$, $\phi$ satisfies

$$\phi_{rr} + \frac{1}{r}\phi_r + \frac{1}{r^2}\phi_{\theta\theta} = 0 \tag{1.11}$$

with the boundary condition

$$\frac{1}{r}\phi_\theta = 0 \quad on\ \theta = 0\ and\ \theta = 2\pi. \tag{1.12}$$

In a small neighbourhood of the origin let

$$\phi = r^\gamma f(\theta) \tag{1.13}$$

where $\gamma$ is to be determined and $f(\theta)$ is a twice differentiable function. Then from (1.11), f($\theta$) must satisfy

$$f''(\theta) + \gamma^2 f(\theta) = 0$$

the solution of which is

$$f(\theta) = A\ cos\gamma\theta + B sin\gamma\theta$$

so that

$$\phi = r^\gamma (A\ cos\gamma\theta + B\ sin\gamma\ \theta).$$

Applying the boundary condition (1.12) we obtain

$$B = 0, \quad \gamma = n/2$$

where n is an integer. In order to ensure that energy must be bounded everywhere including the origin, we must choose n to be the smallest positive number. This asserts that $\gamma = 1/2$ so that

$$\phi = A\, r^{1/2} cos(\theta/2) \quad as\ r \to 0.$$

In the problems of water wave radiation or diffraction involving a body or bodies having submerged sharp edge(s), the velocity potential $\phi$ near a sharp edge thus satisfies

$$\phi_r \sim O(r^{-\frac{1}{2}}) \quad as\ r \to 0, \tag{1.14}$$

r being the distance from the edge. This condition ensures that the energy is bounded in any region including a sharp edge.

# 2 Water wave diffraction by nearly vertical barrier

The problem of water wave diffraction involving obstacles of various geometrical shapes constitutes an important class of problems in the linearised theory of water waves. However, as mentioned earlier the closed form solutions to the problems in this class are available in the literature only when the obstacle is in the form of a thin vertical plane and also for the two-dimensional motion in deep water and for normal incidence of the wave train.

The water wave diffraction problem involving a nearly vertical barrier was first considered by Shaw (1985) wherein the barrier is partially immersed. He used an integral equation formulation based on Green's integral theorem and employed a perturbation technique to solve the integral equation upto first order. The first order corrections to the reflection and transmission coefficients were evaluated involving somewhat tedious analytical calculations. Later Mandal and Chakrabarti (1989) used a simplified perturbation analysis applied to the governing partial differential equation and the boundary conditions, to reduce the basic boundary value problem describing the physical problem considered by Shaw (1985) to two independent boundary value problems upto first order, and deduced Shaw's results for the first order corrections to the reflection and transmission coefficients by using a tailored version of Green's integral theorem without solving the BVP explicitly. The complementary problem of a submerged nearly vertical barrier extending infinitely downwards was also considered by Mandal and Chakrabarti (1989) by employing the same technique and the corresponding problem involving a submerged nearly vertical plate was considered by Mandal and Kundu (1990) by employing Shaw's integral equation technique and also the simplified perturbation expansion method.

In this section we present a brief account of the simplified perturbation technique and the method of evaluation of the first order corrections to the reflection and transmission coefficients, concerning the aforesaid three problems of water wave diffraction involving a nearly vertical thin barrier.

## 2.1 Formulation of the problem

Let Re $\{\phi(x,y)e^{-i\sigma t}\}$ represent the velocity potential for the two-dimensional motion due to the incidence of a time-harmonic progressive wave train described by Re $\{\phi^{inc}(x,y)e^{-i\sigma t}\}$, where $\phi^{inc}$ (x,y) $= e^{-Ky+iKx}$, on a nearly vertical thin barrier. The barrier occupies the position x $= \epsilon c(y)$, y $\epsilon$ L, where $\epsilon$ is a dimensionless small parameter, c(y) is a continuous bounded function of y vanishing at the ends of L and L = (0,a) if the barrier is partially immersed upto a depth 'a' below the mean free surface, L = (b, $\infty$) if it is completely submerged and extends infinitely downwards from a depth 'b' below the mean free surface and L = (a,b) if it is submerged in deep water from a depth 'a' to 'b'. Then $\phi$ satisfies the boundary value problem (BVP) described by the PDE (1.6) and the conditions (1.7), (1.9), (1.10) together with (1.8) whose form now is

$$\frac{\delta\phi}{\delta n} = 0 \ \ on\ x = \epsilon\ c(y), \ \ y\epsilon L. \tag{2.1}$$

Also, since the barrier has submerged sharp edge(s), $\phi$ must satisfy the edge condition given by

$$r^{1/2}\nabla\phi \ \ is\ bounded\ as\ r \ \to\ 0 \tag{2.2}$$

where r denotes the distance(s) from the sharp edge(s) of the barrier.

The condition (2.1) can be approximated to the first order of $\epsilon$ as

$$\phi_x \ - \ \epsilon\frac{d}{dy}\{c(y)\phi_y\} + O(\epsilon^2) \ = \ 0\ on\ x \ = \ \pm 0, y\epsilon L. \tag{2.3}$$

This form of boundary condition suggests the following perturbation expansion in terms of the powers of $\epsilon$ of $\phi$ and the unknown constants R and T

$$\left.\begin{aligned} \phi(x,y,\epsilon) &= \phi_0(x,y) + \epsilon\phi_1(x,y) + O(\epsilon^2), \\ R(\epsilon) &= R_0 + \epsilon R_1 + O(\epsilon^2), \\ T(\epsilon) &= T_0 + \epsilon T_1 + O(\epsilon^2). \end{aligned}\right\} \tag{2.4}$$

Substituting (2.4) in (1.6), (1.7), (2.3), (2.2), (1.9) and (1.10) and equating the coefficients of similar powers of $\epsilon$ from both sides, we find that $\phi_0$ and $\phi_1$ are the solutions of the following two independent BVPs.

*BVP I* $\phi_0$ satisfies

$$\nabla^2\phi_0 \ = \ 0 \ \ in\ y \ \geq \ 0,$$

$$K\phi_0 + \phi_{0y} = 0 \ on \ y = 0,$$
$$\phi_{0x} = 0 \ on \ x = \pm 0, \ y \epsilon L,$$
$$r^{1/2} \nabla \phi_0 \ is \ bounded \ as \ r \to 0$$

where r is the distance from (0,a) or (0,b) or (0,a) and (0,b) according as L = (0,a) or (0,b) or (a,b),

$$\nabla \phi_0 \to 0 \ as \ y \to \infty$$

and

$$\phi_0(x,y) \to \begin{cases} \phi^{inc}(x,y) + R_0\phi^{inc}(-x,y) \ as \ x \to -\infty \\ T_0\phi^{inc}(x,y) \qquad as \ x \to \infty; \end{cases}$$

*BVP II* $\phi_1$ satisfies

$$\nabla^2\phi_1 = 0 \ in \ y \geq 0, \tag{2.5}$$

$$K\phi_1 + \phi_{1y} = 0 \ on \ y = 0, \tag{2.6}$$

$$\phi_{1x}(\pm 0, y) = \frac{d}{dy}\{c(y)\ \phi_{0y}(\pm 0, y)\}, \ y \ \epsilon \ L, \tag{2.7}$$

$$r^{1/2}\nabla\phi_1 \ is \ bounded \ as \ r \to 0 \tag{2.8}$$

where r is as in BVP I,

$$\nabla\phi_1 \to 0 \ as \ y \to \infty \tag{2.9}$$

and

$$\phi_1 \to \begin{cases} R_1\phi^{inc}(-x,y) \ as \ x \to -\infty, \\ T_1\phi^{inc}(x,y) \quad as \ x \to \infty. \end{cases} \tag{2.10}$$

## 2.2 Solution to BVP I

The BVP I corresponds to the problem of water wave diffraction by a plane vertical thin barrier which is partially immersed when L = (0,a), or is completely submerged and extends infinitely downwards when L = (b, $\infty$), or is in the form of a completely submerged plate when L = (a,b). The exact solutions for these three basic configurations of the barrier are well-known in the literature and are given below.

(i) L = (0,a), *partially immersed plate* (cf. Ursell (1947))

$$\phi_0^{(1)}(x,y) = \begin{cases} e^{-Ky+iKx} + R_0^{(1)}e^{-Ky-iKx} + \frac{1}{\Delta_1}\int_0^\infty \frac{J_1(ka)e^{kx}M(k,y)}{k^2+K^2}dk, x < 0, \\ T_0^{(1)}e^{-Ky+iKx} - \frac{1}{\Delta_1}\int_0^\infty \frac{J_1(ka)e^{-kx}M(k,y)}{k^2+K^2}dk, x > 0, \end{cases} \tag{2.11}$$

where

$$\begin{aligned} R_0^{(1)} &= \pi I_1(Ka)\,/\Delta_1, T_0^{(1)} \;=\; iK_1(Ka)\,/\Delta_1 \\ \Delta_1 &= \pi\ I_1(Ka) + iK_1(Ka), \\ M(k,y) &= k\ cosky - Ksinky. \end{aligned} \tag{2.12}$$

(ii) L = (b, $\infty$), *completely submerged barrier extending infinitely downwards* (cf. Ursell (1947), Mandal and Kundu (1987))

$$\phi_0^{(2)}(x,y) = \begin{cases} e^{-Ky+iKx} + R_0^{(2)}e^{-Ky-iKx} - \frac{1}{\Delta_2}\int_0^\infty \frac{J_0(kb)e^{kx}M(k,y)}{k^2+K^2}dk, x<0, \\ T_0^{(2)}e^{-Ky+iKx} + \frac{1}{\Delta_2}\int_0^\infty \frac{J_0(kb)e^{-kx}M(k,y)}{k^2+K^2}dk, x>0, \end{cases} \tag{2.13}$$

where

$$\begin{aligned} R_0^{(2)} &= K_0(Kb)\,/\Delta_2, T_0^{(2)} \;=\; i\pi I_0(Kb)\,/\Delta_2 \\ \Delta_2 &= K_0(Kb) + i\pi\ I_0(Kb), \end{aligned}$$

(iii) L = (a,b), *submerged vertical plate* (cf. Evans (1970), Mandal and Kundu (1990))

$$\phi_0^{(3)}(x,y) = \begin{cases} e^{-Ky+iKx} + R_0^{(3)}e^{-Ky-iKx} + \frac{2i}{\pi\Delta_3}\int_0^\infty \frac{J(k)e^{kx}M(k,y)}{k^2+K^2}dk, x<0, \\ T_0^{(3)}e^{-Ky+iKx} - \frac{2i}{\pi\Delta_3}\int_0^\infty \frac{J(k)e^{-kx}M(k,y)}{k^2+K^2}dk, x>0, \end{cases} \tag{2.14}$$

where

$$R_0^{(3)} = -i\gamma_0/\Delta_3,\quad T_0^{(3)} = \frac{\alpha_0-\beta_0}{\Delta_3}, \Delta_3 = \alpha_0-\beta_0-i\gamma_0,$$

$$\alpha_0 = \int_{-a}^{a}\frac{(d_0^2-u^2)e^{-Ku}}{\{(a^2-u^2)(b^2-u^2)\}^{\frac{1}{2}}}du,\ \beta_0 = \int_b^\infty \frac{(d_0^2-u^2)e^{-Ku}}{\{(u^2-a^2)(u^2-b^2)\}^{\frac{1}{2}}}du,$$

$$\gamma_0 = \int_a^b \frac{(d_0^2-u^2)e^{-Ku}}{\{(u^2-a^2)(b^2-u^2)\}^{\frac{1}{2}}}du,\ J(k) = \int_a^b \frac{(d_0^2-u^2)\sin ku}{\{(u^2-a^2)(b^2-u^2)\}^{\frac{1}{2}}}du,$$

where $d_0^2$ is given by

$$\int_a^b \frac{(d_0^2-u^2)e^{Ku}}{\{(u^2-a^2)(b^2-u^2)\}^{\frac{1}{2}}}du = 0. \tag{2.15}$$

## 2.3 Evaluation of $T_1$ and $R_1$

The BVP II is of some special type as it involves two different boundary conditions on the two sides of a vertical line as is evident from (2.7). Solutions to this problem for the aforesaid three configurations of L will be considered in Section 3. As the first order corrections $R_1$ and $T_1$ of the reflection and transmission coefficients occur in the BVP II only, one may think that these can be evaluated only after solving the BVP II. However, this is not true, and $T_1$ and $R_1$ can be obtained for each of the three cases by utilizing Evans' (1976) idea of the use of a tailored version of Green's integral theorem without solving the BVP II. We now indicate briefly how $R_1$ and $T_1$ can be evaluated for L = (0,a). The details of the calculations for this case as well as for L = (b, $\infty$) and L = (a,b) are given in Mandal and Chakrabarti (1989) and Mandal and Kundu (1990).

For L = (0,a) we apply Green's integral theorem to the functions $\phi_0^{(1)}$ (x,y) and $\phi_1^{(1)}$ (x,y) (the superscript denotes the case L = (0,a)) in the region bounded by the lines

$$y = 0, (0 < x \leq X); x = X, (0 < y \leq Y); y = Y, (-X \leq x \leq X);$$
$$x = -X, (0 \leq y \leq Y); y = 0, (-X \leq x < 0); x = 0-, (0 \leq y \leq a);$$
$$x = 0+, \quad (0 \leq y \leq a)$$

with X, Y > 0 and X,Y are ultimately allowed to tend to infinity. Using the argument similar to that of Evans (1976) we find that

$$iR_1^{(1)} = \int_0^a \left\{\phi_0^{(1)}(+0,y)\phi_{1x}^{(1)}(+0,y) - \phi_0^{(1)}(-0,y)\phi_{1x}^{(1)}(-0,y)\right\} dy \tag{2.16}$$

in which, by virtue of (2.7), we use

$$\phi_{1x}^{(1)}(\pm 0,y) = \frac{d}{dy}\left\{c(y)\phi_{oy}^{(1)}(\pm 0,y)\right\}, 0 < y < a$$

where (cf. Mandal and Chakrabarti (1989))

$$\phi_0^{(1)}(\pm 0,y) = \begin{cases} e^{-Ky} \pm \frac{e^{-Ky}}{\Delta_1 a}\int_a^y \int \frac{te^{Kt}}{(a^2-t^2)^{\frac{1}{2}}} dt \ for \ y < a \\ e^{-Ky} \ for \ y \ \geq \ a. \end{cases} \tag{2.17}$$

Ultimately we find

$$\frac{i\Delta_1 a}{4K} R_1^{(1)} = -K\int_0^a c(y)e^{-2Ky}\left\{\int_a^y \frac{te^{Kt}}{(a^2-t^2)^{\frac{1}{2}}}dt\right\} dy + \int_0^a \frac{c(y)e^{-Ky}}{(a^2-y^2)^{\frac{1}{2}}}dy \tag{2.18}$$

This result can be identified with the result of Shaw (1985) obtained by a different method.

In order to evaluate the correction $T_1^{(1)}$ to the transmission coefficient, we apply the Green's integral theorem to the functions $\psi_0^{(1)}$ (x,y) $\equiv \phi_0^{(1)}$ (-x,y) and $\phi_1^{(1)}$ (x,y) in the same region mentioned above. Here $\psi_0^{(1)}$ (x,y) is the velocity potential for the problem of diffraction of a water wave train normally incident on a partially immersed thin vertical barrier from positive infinity. We obtain in this case

$$iT_1^{(1)} = \int_0^a \left\{\psi_0^{(1)}(+0,y)\ \phi_{1x}^{(1)}(+0,y) - \psi_0^{(1)}(-0,y)\phi_{1x}^{(1)}(-0,y)\right\} dy$$

$$= \int_0^a [\phi_0^{(1)}(-0,y)\frac{d}{dy}\left\{c(y)\phi_{0y}^{(1)}(+0,y)\right\} - \phi_0^{(1)}(+0,y)\frac{d}{dy}\left\{c(y)\phi_{0y}^{(1)}(-0,y)\right\}]dy.$$

Integrating by parts and using the fact that c(0) = 0, this gives

$$T_1^{(1)} = -i[\phi_{0y}^{(1)}(+0,y)\phi_0^{(1)}(-0,y) - \phi_{0y}^{(1)}(-0,y)\phi_0^{(1)}(+0,y)]_{y\ =\ a}$$

Using (2.17) this produces

$$T_1^{(1)} = 0. \tag{2.19}$$

This result was however guessed by Shaw (1985) who used a physical argument based on the fact that the transmission coefficient T remains unaltered if the scattering body is reversed but the incident field is left unaltered. Now if $\epsilon c(y)$ is replaced by $-\epsilon c(y)$, the shape of the scattering body is reversed. As T remains unaltered we must have

$$T = T_0 + \epsilon T_1 + O(\epsilon^2) = T_0 - \epsilon T_1 + O(\epsilon^2)$$

so that $T_1 = 0$. However, Shaw (1985) could not prove this result analytically while for L = (0,a), this is proved here analytically and for L = (b, $\infty$), (a,b). This result is also proved in Mandal and Chakrabarti (1989) and Mandal and Kundu (1990).

For L = (b, $\infty$), the first order corrections $R_1^{(2)}$ and $T_1^{(2)}$ to the reflection and transmission coefficients can be obtained in a similar manner (cf. Mandal and Chakrabarti (1989)) and are given by

$$\frac{i\Delta_2}{4K}R_1^{(2)} = K\int_b^\infty c(y)e^{-2Ky}\left\{\int_b^y \frac{e^{Kt}}{(t^2-b^2)^{\frac{1}{2}}}dt\right\}dy - \int_b^\infty \frac{c(y)e^{-Ky}}{(y^2-b^2)^{\frac{1}{2}}}dy, \tag{2.20}$$

$$T_1^{(2)} = 0. \tag{2.21}$$

For L = (a,b) these quantities are given by (cf. Mandal and Kundu (1990))

$$\frac{\Delta_3}{4K}R_1^{(3)} = K\int_a^b c(y)e^{-2Ky}\left\{\int_y^b \frac{(d_0^2-t^2)e^{-Kt}}{\{(t^2-a^2)(b^2-t^2)\}^{\frac{1}{2}}}dt\right\}dy + \int_a^b \frac{c(y)(d_0^2-y^2)e^{-Ky}}{\{(y^2-a^2)(b^2-y^2)\}^{\frac{1}{2}}}dy, \tag{2.22}$$

$$T_1^{(3)} = 0. \tag{2.23}$$

# 3 Some special boundary value problems arising in linear water wave theory

In this section explicit solutions to boundary value problems of a special type arising in connection with problems of water wave diffraction by a nearly vertical barrier, as mentioned in Section 2, are obtained. These BVPs are of a special type since they involve two different conditions on the two sides of a vertical line. Viewed independently, these may be regarded as problems of water wave radiation by a thin vertical plane on two sides of which two different normal velocities are prescribed. This type of BVP is something new in the water wave theory and cannot perhaps be visualised in general. However, different boundary conditions on two sides of a plane commonly occur in the electromagnetic theory involving half plane diffraction problems.

Havelock's expansion of water wave potential can be utilized to reduce the three BVPs of special type corresponding to L = (0,a), (b, $\infty$) and (a,b) (see Section 2) to appropriate singular integral equations whose solutions can then be used to evaluate the physical quantities $R_1$ and $T_1$. In fact the cases L = (0,a) and (b, $\infty$) were considered by Mandal and Banerjea (1992a, 1993) by this method. Of course, these problems can also be solved by other methods such as the complex variable technique leading to solving appropriate Riemann Hilbert BVPs (cf. Vijayabharati and Chakrabarti (1991) for L = (b, $\infty$) and Vijayabharati et al (1993) for L = (a,b)).

## 3.1 Statement of the BVP

The BVP II of Section 2.1 is restated here. The problem is to find $\phi(x, y)$ which satisfies

$$\nabla^2\phi = 0 \quad in\ y \geq 0, \tag{3.1}$$

$$K\phi + \phi_y = 0 \quad on\ y = 0, \tag{3.2}$$

$$\phi_x\,(\pm\, 0, y) = f_\pm\,(y),\ y \in L, \tag{3.3}$$

$$r^{1/2}\ \nabla\phi\ is\ bounded\ as\ r \rightarrow 0 \tag{3.4}$$

where r is as in BVP I of Section 2.1,

$$\nabla\,\phi \rightarrow 0\ as\ y \rightarrow \infty \tag{3.5}$$

and

$$\phi(x,y) \rightarrow \begin{cases} R_1\ e^{-Ky-iKx}\ as\ x \rightarrow -\infty, \\ T_1\ e^{-Ky+iKx}\ as\ x \rightarrow \infty. \end{cases} \tag{3.6}$$

Here

$$f_{\pm}(y) = \frac{d}{dy}\{c(y)\phi_{0y}(\pm 0, y)\},\ y \in L \tag{3.7}$$

where $\phi_0$ (x,y) is given in Section 2.2 for L = (0,a), L = (b, $\infty$), L = (a,b). In (3.7) we require $\phi_0$ ($\pm$ 0,y). For L = (0,a) (cf. Mandal and Chakrabarti (1989))

$$\phi_0(\pm 0, y) = \begin{cases} e^{-Ky}[1 \pm \frac{1}{\Delta_1 a}\int_a^y \frac{te^{Kt}}{(a^2-t^2)^{1/2}}dt] & for\ y < a, \\ e^{-Ky} & for\ y \geq a; \end{cases} \tag{3.8}$$

for L = (b, $\infty$) (cf. Mandal and Chakrabarti (1989))

$$\phi_0(\pm 0, y) = \begin{cases} e^{-Ky} & for\ y \leq b, \\ e^{-Ky}[1 \mp \frac{1}{\Delta_2}\int_b^y \frac{e^{Kt}}{(t^2-b^2)^{1/2}}dt] & for\ y > b, \end{cases} \tag{3.9}$$

and for L = (a,b) (cf. Mandal and Kundu (1990))

$$\phi_0(\pm 0, y) = \begin{cases} e^{-Ky}\ for\ y \leq a\ and\ b \leq y < \infty, \\ e^{-Ky}[1 \pm \frac{i}{\Delta_3}\int_b^y \frac{(d_0^2-u^2)e^{-ku}}{\{(u^2-a^2)(b^2-u^2)\}^{1/2}}du]\ for\ a < y < b, \end{cases} \tag{3.10}$$

where the constants $\Delta_1$, $\Delta_2$, $\Delta_3$ and $d_0^2$ are given in Section 2.2.

## 3.2 Solution of the BVP

Using Havelock's expansion of water wave potential, a representation of $\phi$ (x,y) satisfying (3.1), (3.2), (3.5) and (3.6) is given by

$$\phi(x, y) = \begin{cases} T_1\ e^{-Ky+iKx} + \int_0^\infty A(k)M(k, y)e^{-kx}dk & for\ x > 0, \\ R_1\ e^{-Ky-iKx} + \int_0^\infty B(k)M(k, y)e^{kx}dk & for\ x < 0 \end{cases} \tag{3.11}$$

where M(k,y) is given in (2.12). If we represent

$$\phi_x(\pm 0, y) = F_{\pm}(y),\ 0 < y < \infty \tag{3.12}$$

then from (3.3) it is evident that

$$F_{\pm}(y) = \begin{cases} f_{\pm}(y) & for\ y \in L \\ f(y) & for\ y \in L' = (0, \infty) - L \end{cases} \tag{3.13}$$

where f(y) is an unknown function and because of (3.4) it has the behaviour

$$f(y) = \begin{cases} O(|\, y-a\,|^{-\frac{1}{2}}) & as\ y \to a\ for\ L = (0, a), \\ O(|\, b-y\,|^{-\frac{1}{2}}) & as\ y \to b\ for\ L = (b, \infty), \end{cases} \tag{3.14}$$

and this has the same combined behaviour as y $\rightarrow$ a and y $\rightarrow$ b for L = (a,b).

Using (3.10) in (3.11) we find

$$iK\begin{bmatrix} T_1 \\ -R_1 \end{bmatrix} e^{-Ky} + \int_0^\infty k \begin{bmatrix} -A(k) \\ B(k) \end{bmatrix} M(k,y)dk = \begin{bmatrix} F_+(y) \\ F_-(y) \end{bmatrix}, y > 0.$$

Utilizing Havelock's theorem, this gives

$$\frac{i}{2}\begin{bmatrix} T_1 \\ -R_1 \end{bmatrix} = \int_{L'} f(y)e^{-Ky}dy + \int_L \begin{bmatrix} f_+(y) \\ f_-(y) \end{bmatrix} e^{-Ky}dy, \tag{3.15}$$

$$\frac{\pi}{2}k(k^2 + K^2)\begin{bmatrix} -A(k) \\ B(k) \end{bmatrix} = \int_{L'} f(y)M(k,y)dy + \int_L \begin{bmatrix} f_+(y) \\ f_-(y) \end{bmatrix} M(k,y)dy. \tag{3.16}$$

Thus from (3.15) we find

$$\frac{i(T_1 + R_1)}{2} = \int_L \{f_+(y) - f_-(y)\} e^{-Ky}dy \tag{3.17}$$

where the right-side can be evaluated in closed form for any of the three forms of L because of (3.7) and (3.8) to (3.10). Thus if we take $T_1 = 0$ on physical considerations as explained by Shaw (1985), then $R_1$ is obtained from (3.17) without actually solving the BVP explicitly (also see Mandal (1991)). Also this leads to some integral identities involving the shape function c(y) for each of the three forms of L.

Now to obtain an integral equation for f(y), we utilize the fact that $\phi$ (x,y) is continuous across the line x = 0, y $\epsilon$ $L'$. Utilizing this in the representation for $\phi$ (x,y) given in (3.11) we find

$$T_1e^{-Ky} + \int_0^\infty A(k)M(k,y)dk = R_1e^{-Ky} + \int_0^\infty B(k)M(k,y)dk \ \ for\ y \in L'. \tag{3.18}$$

Substituting for A(k), B(k) from (3.16), this produces

$$\int_{L'} f(u)\, S(u,y)du = -\frac{1}{2}\int_L \{f_+(u) + f_-(u)\}\, S(u,y)du + \frac{\pi}{2}(T_1 - R_1)e^{-Ky} \ \ \text{for } y \in L' \tag{3.19}$$

where

$$S(u,y) = 2\int_0^\infty \frac{M(k,u)M(k,y)}{k(k^2 + K^2)}dk \tag{3.20}$$

Now applying the operator $(\frac{d}{dy} + K)$ to both sides of (3.19) we obtain a singular integral equation of first kind for f(y) as

$$\int_{L'} f(u)\left\{K\, ln \mid \frac{y-u}{y+u} \mid - \frac{1}{y-u} - \frac{1}{y+u}\right\} du = 2y\mu(y),\ y \in L' \tag{3.21}$$

where $\mu$ (y) involves the shape function and is given by

$$\mu(y) = K \int_L \frac{c(u)e^{-Ku}}{y^2 - u^2} du, y \in L' \tag{3.22}$$

and f(y) satisfies the requirement given in (3.14). For further analysis we consider the cases L = (0,a), L = (b, $\infty$) and L = (a,b) separately.

*Case (i) L = (0,a)*

In this case, L′ = (a, $\infty$). By defining

$$\lambda(u) = f(u) + K \int_a^u f(v)dv, \quad a < u < \infty \tag{3.23}$$

we find that the integral equation (3.21) reduces to the singular integral equation of first kind with a Cauchy type kernel given by

$$\int_a^\infty \frac{\lambda(u)}{y^2 - u^2} du = \mu(y), \quad a < y < \infty \tag{3.24}$$

where, because of (3.14), $\lambda$(y) behaves as

$$\lambda(y) = O(| y - a |^{-1/2}) \ as \ y \to a. \tag{3.25}$$

The solution of (3.24) satisfying (3.25) is given by (cf. Ursell (1947))

$$\lambda(y) = (y^2 - a^2)^{-1/2} \left\{ c_0 y + \frac{2K}{\pi} m(y) \right\} \tag{3.26}$$

where $c_0$ is an arbitrary constant and

$$m(y) = \int_0^a c(t)e^{-Kt} \left\{ \frac{(a^2 - t^2)^{1/2}}{y^2 - t^2} - \frac{a}{y^2} \right\} \frac{dt}{t^2}. \tag{3.27}$$

Then noting (3.23), f(y) is obtained as

$$f(y) = c_0 f_0(y) + \frac{2k}{\pi} f_1(y) \tag{3.28}$$

where

$$f_0(y) = \frac{d}{dy} \left\{ e^{-Ky} \int_a^y \frac{ue^{Ku}}{(u^2 - a^2)^{1/2}} du \right\}, \tag{3.29}$$

$$f_1(y) = \frac{d}{dy} \left\{ e^{-Ky} \int_a^u \frac{u^3 m(u) e^{Ku}}{(u^2 - a^2)^{1/2}} du \right\}.$$

To evaluate the arbitrary constant $c_0$, we substitute f(y) from (3.28) into (3.19). Some manipulations are necessary to evaluate the various integrals, which are given in some detail in Mandal and Banerjea (1992a). After simplification, it is found that the left-side (and also the first term on the right-side) of (3.19) has the factor $e^{-Ky}$ as y - dependence so that an equation for $c_0$ in terms of $(T_1 - R_1)$ is obtained. This is given by

$$\frac{T_1 - R_1}{2} = c_0 \pi a I_1(Ka) + \frac{2}{\pi} KU \tag{3.30}$$

where

$$U = \int_0^a c'(t) e^{-Kt} [2(a^2 - t^2)^{1/2} \int_0^a \frac{u^3 \; sinh^3 Ku}{(a^2 - u^2)^{1/2}} \frac{du}{u^2 - t^2} - a^2 \pi I_1(Ka)] \frac{dt}{t^2}, \tag{3.31}$$

the u-integral being in the sense of Cauchy principal value.

Again, from (3.15) we obtain

$$\frac{i(T_1 - R_1)}{2} = \int_0^a \{f_+(y) + f_-(y)\} e^{-Ky} dy \; + \; 2 \int_a^\infty f(y) e^{-Ky} dy. \tag{3.32}$$

Substituting for $f \pm (y)$ from (3.7) (using (3.8)) and f(y) from (3.28) and after simplifications we find

$$-\frac{T_1 \; - \; R_1}{2} = i c_0 a K_1(Ka) \; + \; \frac{2i}{\pi} a (V - \frac{\pi}{2} \alpha) \tag{3.33}$$

where

$$V = \int_0^a c'(t) e^{-Kt} \left\{ (a^2 - t^2)^{1/2} \int_a^\infty \frac{u^3 e^{-Ku}}{(u^2 - a^2)^{1/2}} \frac{du}{u^2 - t^2} - a^2 K_1(Ka) \right\} \frac{dt}{t^2}, \tag{3.34}$$

$$\alpha = \int_0^a c'(t) e^{-Kt} dt. \tag{3.35}$$

(3.30) and (3.33) now determine $c_0$ as

$$c_0 = -\frac{2K}{\pi a \Delta_1} \left\{ U \; + \; i(V - \frac{\pi}{2}\alpha) \right\}. \tag{3.36}$$

Hence from (3.30) or (3.33) we find

$$\frac{i(T_1 - R_1)\Delta_1}{4K} = I_1(Ka)(V - \frac{\pi}{2}\alpha) - \frac{K_1(Ka)}{\pi} U. \tag{3.37}$$

Again, substituting $f_\pm$ (y) from (3.7) in (3.18) we find after some simplifications

$$\frac{i T_1 \Delta_1 a}{2K} = \int_0^a c'(t) e^{-Kt} N(Ka, t) dt \tag{3.38}$$

where

$$N(Ka,t) = a(a^2 - t^2)^{1/2}[I_1(Ka)\int_a^\infty \frac{ue^{-Ku}}{(u^2-a^2)^{1/2}}\frac{du}{u^2-t^2} - \frac{K_1(Ka)}{\pi}\int_{-a}^{a}\frac{ue^{Ku}}{(u^2-a^2)^{1/2}}\frac{du}{u^2-t^2}]$$

$$-\frac{\pi}{2}aI_1(Ka)e^{-Kt} + \int_t^a \frac{ucoshK(u-t)}{(a^2-u^2)^{1/2}}du, \quad 0 < t < a, \tag{3.39}$$

the second integral inside the square bracket being in the sense of Cauchy principal value. As $T_1$ vanishes identically, we obtain the integral identity involving the shape function c(t) as

$$\int_0^a c'(t)e^{-Kt}N(Ka,t)dt = 0. \tag{3.40}$$

This identity can be associated with the identity obtained by Shaw (1985). However, as also noted by Shaw, it is hard to prove it directly.

The explicit evaluation of $\phi(x,y)$ in this case is straightforward from (3.11)

*Case (ii)* $L = (b, \infty)$

In this case, $L' =$ (o,b), and as in case (i), the integral equation (3.21) can be reduced to a singular integral equation of first kind with a Cauchy type kernel in the interval (0,b) whose solution can be obtained from Mikhlin (1964). This solution is then utilized to solve the BVP completely following the same procedure as in case (i). The details are given in Mandal and Banerjea (1993). In this case the expression for $i(T_1 + R_1)$ /4K is

$$\frac{i(T_1+R_1)}{4K} = K\int_b^\infty c(t)e^{-2Kt}\left\{\int_b^t \frac{e^{Ku}}{(u^2-b^2)^{1/2}}du\right\}dt - \int_b^\infty \frac{c(t)e^{-Kt}}{(t^2-b^2)^{1/2}}dt \tag{3.41}$$

which actually gives $R_1$ since $T_1$ vanishes identically. Also the related integral identity in this case is

$$\int_b^\infty c'(t)e^{-Kt}N(Kb,t)dt = 0 \tag{3.42}$$

where now

$$N(Kb,t) = \frac{t}{\pi}(t^2-b^2)^{1/2}[K_0(Kb)\int_{-b}^{b}\frac{e^{-Ku}}{(b^2-u^2)^{1/2}}\frac{du}{t^2-u^2} - \pi I_0(kb)\int_0^{\infty}\frac{\cos Ku}{(b^2+u^2)^{1/2}}\frac{du}{u^2+t^2}]$$

$$-K_0(Kb)\cosh Kt + \int_b^t \frac{\cosh K(u-t)}{(u^2-b^2)^{1/2}}du, \quad t > b. \tag{3.43}$$

$\phi(x,y)$ can now be obtained from (3.11)

*Case (iii)* $L = (a,b)$

In this case $L'$ consists of the double interval (0,a) and (b, $\infty$) and mainly because of this, the integral equation (3.21) cannot be reduced to another singular integral equation of the first kind with Cauchy type kernel, as carried out in cases (i) and (ii). Thus in this case the integral equation (3.21) needs some special attention for its solution. Banerjea and Mandal (1993) obtained its solution by exploiting an idea due to Tricomi (1951). The solution is given by

$$f(u) = \begin{cases} \frac{d}{du}\left\{e^{-Ku}\int_a^u \lambda_1(v)e^{Kv}dv\right\} \; for \; u < a, \\ \frac{d}{du}\left\{e^{-Ku}\int_b^u \lambda_2(v)e^{Kv}dv\right\} \; for \; u > b. \end{cases} \tag{3.44}$$

where

$$\lambda_1(u) = \frac{B_1(d_1^2-u^2)}{R_0(u)} + \frac{2K}{\pi}\left(\frac{a^2-u^2}{b^2-u^2}\right)^{1/2} P(u) \; for \; u < a, \tag{3.45}$$

$$\lambda_2(u) = \frac{-B_1(d_1^2-u^2)}{R_1(u)} + \frac{2K}{\pi}\left(\frac{u^2-a^2}{u^2-b^2}\right)^{1/2} P(u) \; for \; u > b, \tag{3.46}$$

$$R_0(u) = \{(a^2-u^2)(b^2-u^2)\}^{1/2}, R_1(u) = \{(u^2-a^2)(u^2-b^2)\}^{1/2} \tag{3.47}$$

and

$$P(u) = \frac{1}{2}\int_0^a \frac{\mu(x)}{x^2-u^2}\left(\frac{b^2-x^2}{a^2-x^2}\right)^{1/2} dx + \frac{1}{2}\int_b^{\infty} \frac{\mu(x)}{x^2-u^2}\left(\frac{x^2-b^2}{x^2-a^2}\right)^{1/2} dx. \tag{3.48}$$

Using the form of $\mu$ (x) from (3.22), P(u) can be simplified to

$$P(u) = \frac{\pi K}{2}\int_a^b \frac{\nu c'(\nu)}{u^2-\nu^2}\left(\frac{b^2-\nu^2}{\nu^2-a^2}\right)^{1/2} e^{-K\nu}d\nu. \tag{3.49}$$

Now f(u) involves two unknown constants $B_1$ and $d_1^2$. To determine these constants, (3.44) is substituted in (3.19) which after simplification gives

$$\frac{T_1 - R_1}{2} e^{-Ky} = e^{-Ky} \begin{cases} \int_a^b \lambda_3(u) e^{-Ku} du & for\ y < a, \\ \int_a^b \lambda_3(u)(e^{-Ku} - e^{Ku}) du & for\ y > b \end{cases} \tag{3.50}$$

where

$$\lambda_3(u) = \frac{B_1(d_1^2 - u^2)}{R(u)} - \frac{2K}{\pi}(\frac{u^2 - a^2}{b^2 - u^2})^{1/2} P(u), \tag{3.51}$$

$$R(u) = \left\{(u^2 - a^2)(b^2 - u^2)\right\}^{1/2}.$$

Now (3.50) produces

$$\int_a^b \lambda_3(u) e^{-Ku} du = \int_a^b \lambda_3(u) e^{-Ku} du (e^{-Ku} - e^{Ku}) du$$

so that

$$\int_a^b \lambda_3(u) e^{Ku} du = 0. \tag{3.52}$$

This gives a relation between $B_1$ and $d_1^2$.

Again from (3.50) we find

$$\frac{i}{2}(T_1 - R_1) = i\gamma_1 B_1 - \frac{2iK}{\pi}\gamma_2 \tag{3.53}$$

where

$$\gamma_1 = \int_a^b \frac{(d_1^2 - u^2) e^{-Ku}}{R(u)} du, \gamma_2 = \int_a^b P(u)(\frac{u^2 - a^2}{b^2 - u^2})^{1/2} e^{-Ku} du. \tag{3.54}$$

Also, using (3.44) in (3.15) we find

$$\frac{i}{2}(T_1 - R_1) = B_1(\alpha_1 - \beta_1) + \frac{2K}{\pi}(\alpha_2 + \beta_2) - \frac{K}{2}\int_a^b c'(\nu) e^{-2K\nu} d\nu \tag{3.55}$$

where

$$\alpha_1 = \int_{-a}^a \frac{d_1^2 - u^2}{R_0(u)} e^{-Ku} du, \ \beta_1 = \int_b^\infty \frac{d_1^2 - u^2}{R_1(u)} e^{-Ku} du,$$

$$\alpha_2 = \int_{-a}^a P(u)(\frac{a^2 - u^2}{b^2 - u^2})^{1/2} e^{-Ku} du, \ \ \beta_2 = \int_b^\infty P(u)(\frac{u^2 - a^2}{u^2 - b^2})^{1/2} e^{-Ku} du. \tag{3.56}$$

(3.53) and (3.55) give

$$B_1 D_1 + \frac{2K}{\pi} D_2 - \frac{K}{2}\int_a^b c'(u) e^{-2Ku} du = 0 \tag{3.57}$$

where

$$D_1 = \alpha_1 - \beta_1 - i\gamma_1, D_2 = \alpha_2 - \beta_2 + i\gamma_2. \tag{3.58}$$

(3.57) gives another relation between $B_1$ and $d_1^2$. Thus from (3.52) and (3.57) $B_1$ and $d_1^2$ can be obtained in principle.
Again, using (3.7) in (3.15) we find

$$\frac{(T_1 + R_1)\Delta_3}{4K} = \int_a^b \frac{c(u)(d_0^2 - u^2)e^{-Ku}}{R(u)} du - K \int_a^b c(u)e^{-2Ku} \left\{ \int_a^u \frac{(d_0^2 - t^2)e^{-Kt}}{R(t)} dt \right\} du. \tag{3.59}$$

From (3.59) and (3.55) or (3.53) $T_1$ is obtained explicitly. Since $T_1$ vanishes identically, the related identity involving the shape function in this case is

$$\gamma_1 B_1 - \frac{2K}{\pi}\gamma_2 + \frac{2K}{\Delta_3} \int_a^b \frac{e^{-Ku}c(u)(d_0^2 - u^2)}{R(u)} du - \frac{2K^2}{\Delta_3} \int_a^b c(u)e^{-2Ku} \left\{ \int_u^a \frac{(d_0^2 - t^2)e^{Kt}}{R(t)} dt \right\} du = 0, \tag{3.60}$$

which is, however, difficult to establish directly.

$\phi$ (x,y) can now be obtained in closed form from (3.11) as all the quantities there are now known.

# 4 Waves due to rolling of a nearly vertical plate

The problem of generation of water waves due to rolling of a thin vertical plate is amongst the few water wave problems in the linearised theory which admits of explicit solution. Ursell (1948) first considered this problem while studying the motion due to the rolling of a ship which was modelled as a thin vertical plate partially immersed in deep water and executing small rolling oscillations about a horizontal axis through it. He used Havelock's expansion of water wave potential and evaluated the amplitude of radiated waves at infinity. Later Evans (1970, 1976), Mandal (1991) studied the problem of generation of waves due to small oscillations of a thin vertical plate partially immersed or completely submerged in deep water.

Mandal and Banerjea (1991, 1992b, 1994a) recently considered the problems of generation of water waves due to rolling oscillations of a nearly vertical plate. By employing the simplified perturbation analysis of Section 2, they reduced each problem to two independent BVPs upto first order. One of the BVPs corresponds to the problem of generation of water waves due to rolling of a vertical plate whose solution is already known in the literature. The potential function describing the second BVP describes a symmetric motion about the line x = 0. This observation coupled with Havelock's expansion of water wave potential is utilized to obtain the solution of the second BVP in a closed form for each of the cases when the plate is partially immersed or completely submerged. The first order corrections to the amplitude of radiated waves at infinity on the two sides of the plate are evaluated. Detailed description of the mathematical analysis for the problems of generation of water waves due to small oscillations of a nearly vertical plate partially immersed or completely submerged in deep water is given in Mandal and Banerjea (1991, 1992b, 1994a). In this section a brief account of this analysis and various results concerning a partially immersed nearly vertical plate is given.

## 4.1 Formulation of the problem

We consider a thin rigid nearly vertical plate represented by $x = \epsilon\, c(y)$, $y \,\epsilon\, L$ where $\epsilon$ is a small dimensionless parameter, c(y) is a smooth bounded continuous function for $y \,\epsilon\, L$ and vanishes at the submerged ends of the plate. The plate is either partially immersed (L = (0,a)) or completely submerged (L = (a,b)) in deep water occupying the region $y \geq 0$ with $y = 0$ as the mean free surface. The plate is hinged at ($\epsilon$c(s), s) and is forced to perform simple harmonic oscillations of amplitude $\theta = \text{Re}\,\{\theta_0 e^{-i\sigma t}\}$ about its mean vertical position, $\sigma$ being the frequency of oscillation. Assuming the ensuing motion in the fluid to be irrotational, it can be described by a velocity potential Re $\{\phi(x,y)e^{-i\sigma t}\}$ where $\phi$ satisfies the BVP described by (1.1), (1.2), (1.4) and (1.5) together with the condition on the plate given by

$$\frac{\delta\phi}{\delta n} = i\sigma\theta_0(y-s) \;\; on\; x = \epsilon\, c(y),\; y \,\epsilon\, L \tag{4.1}$$

where n is the outward drawn normal to the plate. Also $\phi$ satisfies the edge condition that

$$r^{\frac{1}{2}}\, \nabla\phi \text{ is bounded as } r \rightarrow 0 \tag{4.2}$$

where r is the distance from the submerged edge(s) of the plate.

The boundary condition (4.1) can be expressed approximately to the first order of $\epsilon$ in the form

$$\phi_x - \epsilon\frac{d}{dy}\{c(y)\phi_y\} = i\sigma\theta_0(y-s) \; on\; x = 0\pm,\; y \,\epsilon\, L. \tag{4.3}$$

This form of the boundary condition suggests the following perturbation expansion of $\phi$, A and B as

$$\left.\begin{aligned} \phi(x,y,\epsilon) &= \phi_0(x,y) + \epsilon\phi_1(x,y) + O(\epsilon^2), \\ A(\epsilon) &= A_0 + \epsilon\, A_1 + O(\epsilon^2), \\ B(\epsilon) &= B_0 + \epsilon\, B_1 + O(\epsilon^2). \end{aligned}\right\} \tag{4.4}$$

Here $A_0$, $B_0$ are the amplitudes (complex) at infinity of the wave motion set up due to rolling oscillations of a vertical plate on its two sides and $A_1$, $B_1$ are the first order corrections to the wave amplitudes at large distances from the nearly vertical plate at its two sides. Substituting (4.4) in (1.1), (1.2), (4.3), (1.4) and (1.5) and equating the coefficients of similar powers of $\epsilon$ from both sides of the equations we find that $\phi_0$ and $\phi_1$ satisfy the following two independent BVPs.

**BVP I** The function $\phi_0$ satisfies

$$\nabla^2 \phi_0 = 0 \ in\ y \geq 0,$$

$$K\phi_0 + \phi_{oy} = 0 \ on\ y = 0,$$

$$\phi_{ox} = i\sigma\theta_0 (y - s) \ on\ x = 0,\ y \epsilon L,$$

$$r^{\frac{1}{2}} \nabla\phi_0 \ is\ bounded\ as\ r \to 0$$

where r is the distance from (0,a) for L = (0,a) and from (0,a) and (0,b) for L = (a,b),

$$\nabla \phi_0 \to 0 \ as\ y \to \infty,$$

$$\phi_0(x,y) \to \begin{cases} A_0\, e^{-Ky+iKx} & as\ x \to \infty, \\ B_0\, e^{-Ky-iKx} & as\ x \to -\infty. \end{cases}$$

**BVP II** The function $\phi_1$ satisfies

$$\nabla^2 \phi_1 = 0 \ in\ y \geq 0, \tag{4.5}$$

$$K\,\phi_1 + \phi_{1y} = 0 \ on\ y = 0, \tag{4.6}$$

$$\phi_{1x}(\pm 0, y) = \frac{d}{dy}\left\{c(y)\phi_{0y}(\pm 0, y)\right\}, y \epsilon L \tag{4.7}$$

$$r^{\frac{1}{2}} \nabla\phi \ is\ bounded\ as\ r \to 0 \tag{4.8}$$

where r is as in BVP I,

$$\nabla\phi_1 \to 0 \ as\ y \to \infty, \tag{4.9}$$

$$\phi_1(x,y) \to \begin{cases} A_1\, e^{-Ky+iKx} & as\ x \to \infty, \\ B_1\, e^{-Ky-iKx} & as\ x \to -\infty. \end{cases} \tag{4.10}$$

In the BVP I it may be noted that $\phi_0$ (x,y) is the velocity potential of the motion set up due to small rolling oscillations of a thin vertical plate which is either partially immersed (L = (0,a)) or completely submerged (L = (a,b)) in deep water. Because of the condition on x = 0, $\phi_0$ is odd in x so that $A_0$ = -$B_0$ (cf. Evans (1976)). The solution of BVP I is known in the literature (cf. Ursell (1948) for L = (0,a), Evans (1970), Mandal (1991) for L = (a,b), Banerjea and Mandal (1992) for both L = (0,a) and L = (a,b)). As in Section 2, BVP II is of some special type as it involves two different conditions on the two sides of a vertical line as is evident from the condition (4.7).

Only the partially immersed nearly vertical plate problem (for which L = (0,a)) is discussed here. The solution of the BVP I is given below. The BVP II is solved implicitly by a method based on Havelock's expansion and exploitation of the evenness property of $\phi_1$ (x,y) in x. The first order correction $A_1$ to the wave amplitude at infinity is obtained in a natural manner. However, this correction can also be obtained without solving the BVP II explicitly by employing the Green's integral theorem judiciously as is done in Section 2 in connection with evaluation of first order corrections to the reflection and transmission coefficients in the water wave scattering problems involving nearly vertical barrier.

The solution of the BVP I for L = (0,a) is given by (cf. Mandal and Banerjea (1991), Banerjea and Mandal (1992))

$$\phi_0(x,y) = \begin{cases} A_0\, e^{-Ky+iKx} + \int_0^\infty \chi(k)M(k,y)e^{-kx}dk & for\ x > 0, \\ -A_0\, e^{-Ky-iKx} - \int_0^\infty \chi(k)M(k,y)e^{kx}dk & for\ x < 0, \end{cases} \tag{4.11}$$

where M(k, y) = k cosky - K sinky.

$$\chi(k) = \frac{\sigma_a\theta_0}{k^2+K^2}[(q+ip)aJ_1(ka) - \frac{iKa}{k}J_2(ka) + \frac{i(Ks-1)}{k}\{J_1(ka)\mathbf{H}_0(ka) - \mathbf{H}_1(ka)J_0(ka)\}], \tag{4.12}$$

$$q = \frac{\pi}{\Delta Ka}[\frac{1}{2} + \frac{Ks-1}{Ka}\{I_1(Ka) + \mathbf{L}_1(Ka)\}], \tag{4.13}$$

$$\Delta = K_1^2(Ka) + \pi^2 I_1^2(Ka), \tag{4.14}$$

$$p = \frac{1}{I_1(Ka)}[\frac{1}{2}I_2(Ka) + \frac{1-Ks}{Ka}\{I_1(Ka)\mathbf{L}_0(Ka) - \mathbf{L}_1(Ka)I_0(Ka)\} - \frac{q}{\pi}K_1(Ka)], \tag{4.15}$$

$$A_0 = \sigma\theta_0 q a^2\{\pi I_1(Ka) - iK_1(Ka)\}. \tag{4.16}$$

Now in the condition (4.7) of BVP II we require the knowledge of $\phi_0(\pm 0, y)$, y $\epsilon$ L = (0,a), this is given by (cf. Mandal and Banerjea (1991))

$$\phi_0(\pm 0, y) = \begin{cases} \pm e^{-Ky}\int_y^a e^{Kt}\,\Im(t)dt & for\ 0 < y < a, \\ 0\ \ for\ y \geq 0 \end{cases} \tag{4.17}$$

where

$$\Im(t) = \sigma a\theta_0[(q+ip)\frac{t}{(a^2-t^2)^{\frac{1}{2}}} - \frac{iK}{2a}t(a^2-t^2)^{1/2} + \frac{2i}{a\pi}(1-Ks)t\ \ln\frac{t}{t+(a^2-t^2)^{1/2}}]. \tag{4.18}$$

## 4.2 Solution to the BVP II for L = (0,a)

Explicit solution to the BVP II is obtained here first by noting the evenness property of $\phi_1$ (x,y ) in x and then exploiting this in the Havelock's expansion of $\phi_1$ (x,y) for x > 0.

Due to (4.17) we have

$$\phi_0(+0,y) = -\phi_0(-0,y),$$

the condition (4.7) of BVP II then becomes

$$\phi_{1x}(\pm 0,y) = \pm f_0(y) \;\; for \; 0 < y < a \tag{4.19}$$

where

$$f_0(y) = \frac{d}{dy}\{c(y)\phi_{0y}(+0,y)\}. \tag{4.20}$$

It follows that $\phi_1$ (x,y) is even in x so that in (4.10)

$$A_1 = B_1 \tag{4.21}$$

and also

$$\phi_{1x}(+0,y) = -\phi_{1x}(-0,y) \;\; for \; 0 < y < \infty \tag{4.22}$$

But as $\phi_{1x}$ (x,y) is continuous across x = 0 for a < y < $\infty$, (4.22) gives

$$\phi_{1x}(\pm 0,y) = 0 \;\; for \; a < y < \infty. \tag{4.23}$$

Now by Havelock's expansion of water wave potential, an appropriate representation of $\phi_1$ (x,y) for x > 0 is

$$\left.\begin{array}{rcl} \phi_1(x,y) & = & A_1 e^{-Ky+iKx} + \int_0^\infty C(k)M(k,y)e^{-kx}dk \\ and & & for \; x < 0, \;\; \phi_1(x,y) = \phi_1(-x,y). \end{array}\right\} \tag{4.24}$$

Using (4.19), (4.23) and (4.24) we find

$$iA_1e^{-Ky} - \int_0^\infty kC(k)M(k,y)dk = \begin{cases} f_0(y) \; for \; 0 < y < a, \\ 0 \;\; for \; a < y < \infty. \end{cases}$$

By Havelock's inversion theorem this gives

$$iA_1 = 2\int_0^a f_0(y)e^{-Ky}dy \tag{4.25}$$

and

$$C(k) = -\frac{2}{\pi}\frac{1}{k(k^2+K^2)}\int_0^a f_0(y)M(k,y)dy. \tag{4.26}$$

As $f_0$ (y) is known explicitly from (4.20) and (4.17), $A_1$ and C(k) are obtained as

$$A_1 = 2iK \int_0^a \Im(t) g(t) dt, \tag{4.27}$$

$$C(k) = \frac{2}{\pi} \frac{1}{k^2 + K^2} \int_0^a \Im(t) h(t) dt \tag{4.28}$$

where

$$g(t) = e^{Kt} \int_0^t e^{-2Ku} \frac{d}{du} \left\{ c(u) e^{-Ku} \right\} du, \tag{4.29}$$

$$h(t) = e^{Kt} \int_0^t e^{-Ku} \frac{d}{du} \{ c(u)(k sinku + K cosku) \} du, \tag{4.30}$$

and both are related to the shape of the curved plate.
Finally, $\phi_1$ (x,y) is now obtained explicitly from (4.24) since $A_1$ and C(k) are now known.

As mentioned earlier, $A_1$ can also be obtained without solving the BVP II explicitly. This is demonstrated below. In BVP II, we do not take initially $A_1 = B_1$. However, this will be proved in the sequel. This analysis is given in Mandal and Banerjea (1991). Let $\psi_0$ (x,y) be the potential function for the problem of water wave diffraction by a partially immersed thin vertical barrier due to a normally incident wave train represented by $e^{-Ky+iKx}$. $\psi_0$ (x,y) is given by (2.11) of Section 2 (where the notation $\phi_0^{(1)}$(x,y) was used).

In order to obtain $B_1$ of (4.10), we apply Green's integral theorem to the functions $\phi_1$ (x,y) and $\psi_0$ (x,y) in the region bounded by the lines

$$y = 0, (0 < x \le X); \quad x = X, (0 \le y \le Y); \quad y = Y, (-X \le x \le X);$$

$$x = -X, (0 \le y \le Y); \quad y = 0, (-X \le x < 0); \quad x = 0-, (0 \le y \le a);$$

$$x = 0+, (0 \le y \le a) \;\; with \; X, Y > 0$$

and ultimately make X, Y $\to \infty$. Using the arguments similar to that in Section 2, we find that

$$-iB_1 = \int_0^a \{\psi_0(+0,y)\phi_{1x}(+0,y) - \psi_0(-0,y)\phi_{1x}(-0,y)\} dy.$$

Noting $\psi_0(\pm 0, y)$ from Section 3 (eqn (3.8) where the notation $\phi_0(\pm 0, y)$ has been used), we find

$$B_1 = i \int_0^a e^{-Ky} \frac{d}{dy} [c(y) \frac{d}{dy} \{\phi_0(+0,y) - \phi_0(-0,y)\}] dy + \int_0^a e^{-Ky} N(y) \frac{d}{dy} [c(y) \frac{d}{dy} \{\phi(+0,y) + \phi_0(-0,y)\}] dy, \tag{4.31}$$

where

$$N(y) = \frac{1}{\Delta_1 a}\int_a^y \frac{te^{Kt}}{(a^2-t^2)^{\frac{1}{2}}}dt, \tag{4.32}$$

$\Delta_1$ being defined in Section 2 (eqn (2.11)).
Since $\phi_0$ (+0,y) = - $\phi_0$ (-0,y) (by (4.17)), the second term in the right-side of (4.31) vanishes. Using (4.17), we find finally

$$B_1 = 2iK\int_0^a \Im(t)g(t)dt \tag{4.33}$$

where $\Im$ (t) and g(t) are given by (4.18) and (4.29) respectively.

Again, to evaluate $A_1$, we apply the Green's integral theorem to $\phi_1$ (x,y) and $\psi_0$ (-x,y) in the same region and it is found that

$$A_1 = 2iK\int_0^a \Im(t)g(t)dt = B_1 \tag{4.34}$$

which was already obtained in (4.27) by a different method.

Considering some particular forms of the shape of the curved plate, Mandal and Banerjea (1991) also calculated numerically the amplitude of the wave motion at large distances from the plate on its two sides and compared with the vertical plate results. For the sake of completeness, their related analysis is briefly given here. Assuming a general Fourier expansion for G($\theta$) $\equiv$ $g(asin\theta)/$a ($0\leq \theta \leq \theta/2$) given by

$$G(\theta) = \frac{a_0}{2} + \sum_{n=1}^{\infty}(a_n cos4n\theta + b_n sin4n\theta) \tag{4.35}$$

where $a_n$, $b_n$s can be obtained once the shape of the plate is known, we find

$$A_1 = 2\sigma a^2\theta_0 K[\frac{a_0\alpha_0}{2} + \sum_{n=1}^{\infty}(a_n\alpha_n + b_n\beta_n)], \tag{4.36}$$

where

$$\alpha_0 = -(p-iq) + \frac{Ka}{3} + \frac{Ks-1}{\pi},$$

$$\alpha_n = \left\{p - iq + \frac{Ka}{2}\frac{16n^2+3}{9-16n^2} + \frac{16n(Ks-1)}{\pi}\right\}\frac{1}{16n^2-1},$$

$$\beta_n = \left\{4(\mathrm{p-iq})\mathrm{n} + 2\mathrm{Kan}\frac{5-8\mathrm{n}^2}{16\mathrm{n}^2-9} + \frac{1-\mathrm{Ks}}{\pi(2\mathrm{n}-1)}\right\}\frac{1}{16\mathrm{n}^2-1} + \frac{\mathrm{Ks}-1}{2\pi(1-4\mathrm{n}^2)}\mathrm{S_n}, \tag{4.37}$$

with

$$S_n = \frac{4}{1-16n^2} + S_{n-1}(n = 1, 2, ...), \quad S_0 = 2 \tag{4.38}$$

from which all $S_n$'s can be calculated.

$\alpha_n$ and $\beta_n$ are numerical constants independent of the shape of the curved plate and can be calculated for each n = 1,2, .. once Ka and s/a are known. Some representative values of these constants are given in Table 1 of Mandal and Banerjea (1991). It may be noted that for any fixed values of the parameters Ka and s/a, $\alpha_n$ and $\beta_n$ decrease with the increase of n. Choosing the shape function c(y) as

$$c(y) = a\,(1 - \frac{y}{a})^m \tag{4.39}$$

where m is a positive integer, G($\theta$) is evaluated as

$$G(\theta) = \frac{1}{2}\left\{e^{-Kasin\theta}(1 - sin\theta)^m - e^{Kasin\theta}\right\}$$

$$+\sum_{j=1}^{m} \frac{m(m-1)\cdots(m-j+1)}{(-2)^{j+1}(Ka)^j}\left\{e^{-Kasin\theta}(1 - sin\theta)^{m-j} - e^{Kasin\theta}\right\}. \tag{4.40}$$

The Fourier coefficients $a_n$ (n = 0,1,2,...) and $b_n$ (n = 1,2,...) in the expansion (4.35) for G($\theta$) cannot be found explicitly. However, they can be calculated numerically for given values of Ka and m.

Mandal and Banerjea (1991) evaluated numerically the quantitites $| A_0^* | = | A_0 |/(g\theta_0\ a^{\frac{3}{2}})$, $| A^* | = | A_0 + \epsilon A_1 | \ /\ (g\theta_0\ a^{\frac{3}{2}})$ and $B^* = | -A_0 + \epsilon A_1 | /(g\ \theta_0\ a^{\frac{3}{2}})$ taking $\epsilon$ = .001 and .0005 and for values of Ka between 0.1 and 2.0 and m between 1 and 5. Here $| A_0 |$ is the amplitude (actual) of the wave motion at large distances set up by the rolling oscillations of a partially immersed vertical plate, $| A |$ and $| B |$ are the same (upto first order of $\epsilon$ ) for a curved plate at its right and left sides respectively. From the numerical results, it was observed that $| A_0^* |$ first increases and then decreases as Ka increases from 0.1 and 2.0 for a fixed value of m. Similar behaviour is also observed for $| A^* |$ and $| B^* |$. Thus the qualitative behaviour of the amplitudes at infinity of the wave motion set up due to rolling oscillations of an immersed curved plate is the same as that for a vertical plate with respect to the wave number Ka. Again it was observed that $| A^* |$ and $| B^* |$ for a fixed m differ significantly from $| A_0^* |$ for moderately large values of Ka, and for a fixed Ka and fixed $\epsilon$, $| A^* |$ decreases, while $| B^* |$ increases slowly as m increases from one. This is plausible since the variation of m causes a very slow change in the free surface slope $(= -\ \frac{1}{\epsilon m})$ of the curved plate as $\epsilon$ is much smaller than m. Thus the influence of $\epsilon$ and m on the amplitudes at large distances from the plate is to some extent significant.

# 5 Water wave diffraction by a submerged nearly circular cylinder

The two-dimensional problem of surface water wave diffraction by a long horizontal circular cylinder completely submerged in deep water is well studied in the literature within the framework of linear theory. Dean (1948) used a complex variable technique and Ursell (1950) used the method of multipoles to solve the problem. Oglivie (1963) considered some more general situation and used Ursell's technique to solve it. Levine (1965) used an integral equation formulation based on Green's integral theorem and also considered obliquely incident wave trains. All of them arrived at the remarkable conclusion that a normally incident wave train experiences no reflection. However, if the cross section of the cylinder is not quite circular, then it does experience reflection in general. In this section we consider the case of a nearly circular cylinder and employ the simplified perturbation procedure used in earlier sections to study the problem mathematically and derive expressions for reflection and transmission coefficients correct to order $\epsilon$, where $\epsilon$ is a measure of small departure of the cylinder cross section from circularity. The results presented here are based on a forthcoming paper by the authors (cf. Mandal and Banerjea (1994b)).

## 5.1 Formulation of the problem

A time harmonic surface water wave train represented by Re $\{\phi^{inc}(x,y)e^{-i\sigma t}\}$ is normally incident upon a nearly circular cylinder submerged in deep water from negative infinity. The ensuing motion can be described by a velocity potential Re $\{\phi(x,y)e^{-i\sigma t}\}$ where $\phi$ satisfies the following BVP :

$$\nabla^2\phi = 0 \quad \textit{in the fluid region}, \tag{5.1}$$

$$K\phi + \phi_y = 0 \quad \textit{on } y = -f, \tag{5.2}$$

$$\frac{\delta\phi}{\delta n} = 0 \quad \textit{on } r = a\{1+\epsilon c(\theta)\},\ 0 \leq \theta \leq 2\pi, \tag{5.3}$$

where n is the outward normal to the surface of the cylinder,

$$\nabla\phi \to 0 \quad \textit{as } y \to \infty \tag{5.4}$$

and

$$\phi \to \begin{cases} T\,\phi^{inc}(x,y) & \textit{as } x \to \infty, \\ \phi^{inc}(x,y) + R\,\phi^{inc}(-x,y) & \textit{as } x \to -\infty \end{cases} \tag{5.5}$$

where

$$\phi^{inc}(x,y) = e^{-K(y+f)+iKx}, \tag{5.6}$$

T and R denote the transmission and reflection coefficients respectively and are to be determined.

Here the plane represented by y = -f is the undisturbed free surface, the submerged cylinder is represented by r = a $\{1+\epsilon c(\theta)\}$, $0 \leq \theta \leq 2\pi$ with x = r sin $\theta$, y = r cos $\theta$ where $\epsilon$ is a small non-dimensional parameter and c($\theta$) is a smooth bounded function, f is the depth of the axis of the cylinder below the mean free surface and generators of the cylinder are parallel to the z - axis.

The condition (5.3) can be approximated to the first order of $\epsilon$ as

$$\phi_r - \epsilon[c(\theta)\phi_r + \frac{1}{a}\frac{d}{d\theta}\{c(\theta)\phi_\theta\}] + O(\epsilon^2) = 0 \text{ on } r = a. \tag{5.7}$$

This form of the boundary condition suggests as before

$$\begin{aligned}
\varphi &= \phi_0 + \epsilon\phi_1 + O(\epsilon^2), \\
R &= R_0 + \epsilon R_1 + O(\epsilon^2), \\
T &= T_0 + \epsilon T_1 + O(\epsilon^2).
\end{aligned}$$

Substituting these in the basic partial differential equation and the boundary conditions, and equating the coefficients of the like powers of $\epsilon$ from both sides of the equations, we find that $\phi_0$ and $\phi_1$ satisfy the following two BVPs :

**BVP I** $\phi_0$ (x ,y) satisfies

$$\begin{aligned}
\nabla^2\phi_0 &= 0 \; in \; y > -f, \; r > a, \\
K\phi_0 + \phi_{0y} &= 0 \;\; on \; y = -f, \\
\phi_{0r} &= 0 \;\; on \; r = a, \\
\nabla\phi_0 &\to 0 \; as \; y \to \infty,
\end{aligned}$$

$$\phi_0 \to \begin{cases} \phi^{inc}(x,y) + R_0\,\phi^{inc}(-x,y) & as \; x \to -\infty, \\ T_0\phi^{inc}(x,y) \;\; as \; x \to \infty. \end{cases}$$

**BVP II** $\phi_1$ (x,y) satisfies

$$\nabla^2\phi_1 = 0 \; in \; y > -f, \; r > a, \tag{5.8}$$

$$K\phi_1 + \phi_{1y} = 0 \;\; on \; y = -f, \tag{5.9}$$

$$\phi_{1r} = h(\theta) \;\; on \; r = a, \tag{5.10}$$

where

$$h(\theta) = \frac{1}{a}\frac{d}{d\theta}\left\{c(\theta)\frac{\delta\phi_0}{\delta\theta}(\theta)\right\} \tag{5.11}$$

with $\phi_0(\theta) = \phi_0$ (a sin $\theta$, a cos $\theta$),

$$\nabla\phi_1 \to 0 \;\; as \; y \to \infty, \tag{5.12}$$

$$and \;\; \phi_1 \to \begin{cases} R_1\phi^{inc}(-x,y) & as \; x \to -\infty, \\ T_1\phi^{inc}(x,y) & as \; x \to \infty. \end{cases} \tag{5.13}$$

The BVP I corresponds to the two-dimensional problem of water wave scattering by an infinitely long horizontal circular cylinder submerged in deep water. As stated earlier, this problem is well studied in the literature. Following Levine (1965) $\phi_0(\theta)$ is given by

$$\phi_0(\theta) = \phi^{inc}(\theta) + \sum_{n=0}^{\infty} a_n e^{-in\theta}, 0 \le \theta \le 2\pi \tag{5.14}$$

where $a_n s$ (n $\geq$ 1) satisfy the linear system

$$\sum_{m=1}^{\infty} a_m K_{mn} = (-1)^n \pi \frac{(Ka)^n}{(n-1)!}, n \geq 1 \tag{5.15}$$

with

$$\mathrm{K_{mn}} = \frac{\mathrm{n}\pi}{2}\delta_{\mathrm{mn}} + (-1)^{\mathrm{m+n}}\frac{\pi}{2}\left(\frac{\mathrm{a}}{2\mathrm{f}}\right)^{\mathrm{m+n}}\frac{(\mathrm{m+n-1})!}{(\mathrm{m-1})!(\mathrm{n-1})!} - \frac{\pi(\mathrm{Ka})^{\mathrm{m+n}}}{(\mathrm{m-1})!(\mathrm{n-1})!}\left[\frac{\mathrm{d^{m+n}F}(\mu)}{\mathrm{d}\mu^{\mathrm{m+n}}}\right]\mu = 2\mathrm{Kf},$$

$$K_{0n} = 0 \ \ for \ all \ n \geq 1,$$

$$\delta_{mn} = \begin{cases} 0 \ for \ m \neq n \\ 1 \ for \ m = n \end{cases}$$

and

$$F(\mu) = \int_0^{\infty} \frac{e^{-\mu z}}{z-1} dz \tag{5.16}$$

where the path of integration is indented below the pole at z = 1. It is known that

$$R_0 = 0 \tag{5.17}$$

while

$$T_0 = 1 + 2\pi i e^{-2Kf} \sum_{n=1}^{\infty} a_n \frac{(-1)^n (Ka)^n}{(n-1)!}. \tag{5.18}$$

It is obvious that these results are independent of $a_0$, and $a_n$ (n = 1,2,...) can be determined approximately by truncating the linear system (5.15).

To evaluate $R_1$ we apply Green's integral theorem to $\phi_1$ (x,y) and $\phi_0$ (x,y) in the region bounded by the lines y = -f, (-X $\leq x \leq X$); x = X, (-f $\leq y \leq Y$); y = Y, (-X $\leq x \leq X$); x = -X, (-f $\leq y \leq Y$) and the circle r = a and ultimately make X, Y $\rightarrow$ $\infty$ so that we find

$$R_1 = ia \int_0^{2\pi} \phi_0(\theta) h(\theta) d\theta$$

$$= -i \int_0^{2\pi} \{\phi_{0\theta}(\theta)\}^2 \, c(\theta) d\theta. \tag{5.19}$$

Similarly $T_1$ is obtained by applying Green's integral theorem to $\phi_1$ (x,y) and $\phi_0$ (-x,y) in the same region as

$$T_1 = ia \int_0^{2\pi} \phi_0(-\theta)h(\theta)d\theta$$

$$= -i \int_0^{2\pi} \phi_{0\theta}(\theta)\ \phi_{0\theta}(-\theta)\ c(\theta)d\theta. \tag{5.20}$$

With the knowledge of $\phi_0(\theta)$ given in (5.14) we find

$$R_1 = i \int_0^{2\pi} c(\theta)[(Ka)^2 e^{-2Kf-2i\theta-2Kae^{-i\theta}}$$
$$-2Kae^{-Kf} \sum_{n=1}^{\infty} na_n e^{-i(n+2)\theta-Kae^{-i\theta}}$$
$$+\sum_{m=1}^{\infty}\sum_{n=1}^{\infty} mna_m a_n e^{-i(m+n)\theta}]d\theta, \tag{5.21}$$

and

$$T_1 = i \int_0^{2\pi} c(\theta)[(Ka)^2 e^{-2Kf-2Kacos\theta} + \sum_{m=1}^{\infty}\sum_{n=1}^{\infty} mna_m a_n e^{i(m-n)\theta}$$
$$-2Kae^{-Kf}\sum_{n=1}^{\infty} na_n e^{-Kacos\theta} cos\,\{Kasin\theta + (n-1)\theta\}]d\theta. \tag{5.22}$$

Assuming a general Fourier expansion of $c(\theta)$ given by

$$c(\theta) = \frac{t_0}{2} + \sum_{p=1}^{\infty}(s_p\ sinp\theta + t_p\ cosp\theta), 0 \le \theta \le 2\pi, \tag{5.23}$$

we find

$$\mathrm{R_1} = \pi \sum_{p=2}^{\infty}(s_p + it_p)\left[\frac{(2Ka)^p}{2(p-2)!}e^{-2Kf} + \sum_{n=1}^{p-1} n(p-n)a_n a_{p-n} - 2e^{-Kf}\sum_{n=1}^{p-1}\frac{na_n(Ka)^{p-n}}{(p-n-1)!}\right] \tag{5.24}$$

and

$$\mathrm{T_1} = 2\pi i \sum_{p=1}^{\infty} t_p\left[(Ka)^2 e^{-2Kf} I_p(2Ka) - e^{-Kf}\left\{\sum_{n=p+1}^{\infty}\frac{(Ka)^{n-p}}{(n-p-1)!}na_n + \sum_{n=1}^{\infty}\frac{(Ka)^{n+p}}{(n+p-1)!}na_n\right\}\right]$$

$$+\pi i\left[\sum_{n=1}^{\infty}\sum_{m=n+1}^{\infty} 2mna_m a_n t_{m-n} + t_0\sum_{n=1}^{\infty} n^2a_n^2 + ie^{-2Kf}(Ka)^2 I_0(2Ka)t_0 - 2e^{-Kf}t_0\sum_{n=1}^{\infty}\frac{(Ka)^2}{(n-1)!}na_n\right], \tag{5.25}$$

where $I_p$ (z) denotes the modified Bessel function of first kind.

It may be noted that $R_1$ is independent of $t_0$, $s_1$ and $t_1$, while $T_1$ is independent of $s_p$. Thus if the nearly circular cylinder has the shape for which

$$c(\theta) = \frac{t_0}{2} + s_1 \sin\theta + t_1 \cos\theta, \tag{5.26}$$

then $R_1 = 0$. Also if

$$c(\theta) = \sum_{p=1}^{\infty} s_p \sin p\theta \tag{5.27}$$

then $T_1 = 0$. This implies that for the nearly circular cylinder with shape function $c(\theta)$ given by (5.27) the transmittivity is totally unaffected by its noncircularity.

It may also be observed that if a nearly circular cylinder has sufficient continuity, then the Fourier coefficients $s_j$ and $t_j$ in (5.23) decay rapidly with j and as such a three term series given in (5.26) is a good approximation to many fairly smooth cylinders. This implies that for many smooth nearly circular cylinders, the result that for circular cylinder the reflection coefficient vanishes, is approximately valid.

# References


1. Banerjea, Sudeshna & Mandal, B.N. (1992) On waves due to rolling of a vertical plate, *Indian J. pure appl. Math.*, **23**, 753-761.

2. Banerjea, Sudeshna & Mandal, B.N. (1993) Solution of a singular integral equation in double interval arising in the theory of water waves, *Appl. Math. Lett.*, **6** No. 3, 81-84.

3. Dean, W.R. (1945) On the reflection of surface waves by submerged plane barrier, *Proc. Cam. Phil. Soc.*, **41**, 231-238.

4. Dean, W.R. (1948) On the reflection of surface waves by a submerged circular cylinder, *Proc. Camb. Phil. Soc.*, **44**, 483-491.

5. Evans, D.V. (1970) Diffraction of water waves by a submerged vertical plate, *J. Fluid Mech.*, **40**, 433-451.

6. Evans, D.V. (1976) A note on the waves produced by the small oscillations of a partially immersed vertical plate, *J. Inst. Maths. Applics.* **17**, 135-140.

7. Haskind, M.D. (1959) Radiation and diffraction of surface waves by a flat plate floating vertically, *Prikl. Mat. Mech.* **23**, 546-559 (in Russian).

8. Karp, S.N. & Karal, F.C. (Jr.) (1962) The elastic field behaviour in the neighbourhood of a crack of arbitrary angle, *Comm. Pure Appl. Math.* **15**, 413-421.

9. Levine, H. (1965) Scattering of surface waves by a submerged cylinder, *J. Math. Phys.*, **6**, 1231-1243.

10. Mandal, B.N. (1991) A note on the reflection coefficient for a class of water wave scattering problems, *Appl. Math. Lett.*, **6**, No. 2, 49-52.

11. Mandal, B.N. & Banerjea, S. (1991) A note on waves due to rolling of a partially immersed nearly vertical plate, *SIAM J. Appl. Math.*, **51**, 930-939.

12. Mandal, B.N. & Banerjea, S. (1992a) Solution of a boundary value problem associated with diffraction of water waves by a partially immersed nearly vertical barrier, *ZAMM.*, **72**, 517-519.

13. Mandal, B.N. & Banerjea, Sudeshna (1992b) On waves due to rolling of a submerged nearly vertical plate, *Int. J. Engng. Sci.*, **30**, 341-347.

14. Mandal, B.N. & Banerjea, Sudeshna (1993) Solution of a boundary value problem associated with diffraction of water waves by a submerged nearly vertical barrier, *Bull. Cal. Math. Soc.*, **85**, 209-214.

15. Mandal, B.N. & Banerjea, Sudeshna (1994a) Surface water waves due to rolling oscillations of a nearly vertical plate, *Revu Roum. Sci. Tech. Mec. Appliquee* (to appear).

16. Mandal, B.N. & Banerjea, Sudeshna (1994b) Scattering of water waves by a submerged nearly circular cylinder, *J. Austral. Math. Soc. Ser. B.* (to appear).

17. Mandal, B.N. & Chakrabarti, A. (1989) A note on diffraction of water waves by a nearly vertical barrier, *IMA J. Appl Math.*, **43**, 157-165.

18. Mandal, B.N. & Kundu, P.K. (1987) Scattering of water waves by a vertical barrier and associated mathematical methods, *Proc. Indian natn. Sci. Acad.*, **53A**, 514-530.

19. Mandal, B.N. & Kundu, P.K. (1990) Scattering of water waves by a submerged nearly vertical plate, *SIAM J. Appl. Math.*, **50**, 1221-1231.

20. Mikhlin, S.G. (1964) *Integral Equations*, Pergamon Press, London.

21. Ogilivie, T.F. (1963) First and second order forces on a cylinder submerged under a free surface, *J. Fluid. Mech.*, **16**, 451-472.

22. Segel, L.A. (1977) *Mathematics Applied to Continuum Mechanics* , Macmillan Publishing Co. Inc., New York.

23. Shaw, D.C. (1985) Perturbational results for diffraction of water waves by a nearly vertical barrier, *IMA J. Appl. Math.*, **34**, 99-117.

24. Tricomi, F.G. (1951) The aerofoil equation for double interval, *ZAMP*, **11**, 402-406.

25. Ursell, F. (1947) The effect of a fixed vertical barrier on surface waves in deep water, *Proc. Camb. Phil. Soc.*, **43**, 374-382.

26. Ursell, F. (1948) On waves due to rolling of a ship, *Quart. J. Mech. Appl. Math.*, **1**, 246-252.

27. Ursell, F. (1950) Surface waves in deep water in the presence of a submerged circular cylinder, *Proc. Camb. Phil. Soc.*, **46**, 141-155.

28. Vijayabharati, L. & Chakrabarti, A. (1991) Solution of boundary value problem associated with diffraction of water waves, *IMA J. Appl. Math.*, **47**, 23-32.

29. Vijayabharati, L., Chakrabarti, A., Mandal, B.N. & Banerjea, S. (1993) Solution of the problem of scattering of water waves by a nearly vertical plate, *J. Austral. Math. Soc. Ser. B*.

30. Whitham, G.B. (1974) *Linear and Nonlinear Waves*, Wiley, New York.

# Chapter 4

# Painlevé analysis, similarity reductions and special solutions of nonlinear evolution equations

S. Roy Choudhury, L. Debnath

*Department of Mathematics, University of Central Florida, Orlando, Florida 32816-6990, USA*

## Abstract

Painlevé analysis, stretching transformations, and Lie group analysis are employed to investigate the solutions of the Melnikov equations. The equations are found to possess only the conditional Painlevé property. Non-trivial similarity-reduced solutions resulting from the use of stretching transformations are not reducible to elliptic functions or any of the Painlevé transcendents and, in general, may have movable essential singularities. Only trivial solutions result from the use of Lie group techniques, and from using generalized expansions truncated at various levels. Painlevé expansions, truncated at various levels, are also used to derive new analytical solutions of the Zakharov-Kuznetsov equation. In particular, two families of shock-type solutions, different from known solitary-wave solutions, are obtained. The first, three-parameter family of solutions consists of bidirectional traveling waves with variable phase-speeds along both spatial coordinates. The second, two-parameter family of solutions consists of arbitrary-speed rightward traveling waves along $x$ and bidirectional waves along $y$. The parameters may be chosen to model different sets of numerical or physical data.

## List of symbols

| | |
|---|---|
| $U$ | physical variable-speed $[lt^{-1}]$ |
| $\phi$ | electrostatic potential |
| $\tilde{\phi}$ | location of singular manifold |
| $\psi$ | simplified family of singular manifolds |
| $\alpha, \beta, r, a_i, \beta_i$ | parameters in Painlevé expansions |
| $U_i$ | coefficients in Painlevé expansions |
| $C_i, V, \tilde{V}$ | homographic invariants |
| $\eta, \gamma, \zeta$ | Lie group extensions |
| $\eta^m, \gamma^m, \zeta^m, m = t, x, y$ | extensions in Lie group analysis |

# 1 Introduction

Classically, the method for finding similarity reductions of a given partial differential equation is the Lie group method of infinitesimal transformations.[1] Recent descriptions of this method are contained in Refs 2-4. The method is entirely algorithmic, but often involves significant amounts of algebraic manipulations (see Ref. 5 for a review of the determination of Lie point symmetries using computer algebra programs). A generalization of Lie's method, called the non-classical method of group-invariant solutions, has been developed by Bluman and Cole.[6] This method has, in turn, been generalized by Olver and Rosenau.[7] All of these methods yield Lie point symmetries of a given (family of) partial differential equations.

Noether[8] generalized Lie's method in a different fashion by allowing the transformation to depend upon derivatives of the dependent variable, in addition to the dependent and independent variables. An algorithmic method for determining the resulting 'Lie-Bäcklund' symmetries has been developed.[9] New classes of symmetries, of neither the Lie point or Lie-Bäcklund variety, have also been developed.[10]

All of the above methods involve the use of group theory. Recently, Clarkson and Kruskal[11] have developed a direct method of deriving similarity reductions of partial differential equations. This method yields the most general similarity reductions admitted by an equation and does not employ group theory. This method has recently been generalized by Clarkson and Winternitz[12] for partial differential equations with more than two independent variables.

Although not yet fully proven, the Painlevé tests[13] seem to provide extremely useful necessary conditions for identifying the completely integrable cases of a wide variety of families of nonlinear ordinary and partial differential equations. Originally, Ablowitz et al[14] conjectured that a nonlinear partial differential equation is integrable if <u>all</u> its exact reductions to ordinary differential equations have the Painlevé property. This approach poses the obvious operational difficulty of finding <u>all</u> exact reductions. This difficulty was circumvented by Weiss et al[15] by postulating that a partial differential equation has the Painlevé property if its solutions are single-valued about a movable singular manifold. In this paper, we follow this latter approach to perform the Painlevé analysis of several nonlinear evolution equations.

Our analysis will be in the spirit of other recent work on Painlevé analysis of nonlinear equations.[13,16] There is now a compelling body of evidence that if an equation possesses the Painlevé property it is indeed integrable, i.e. the Painlevé test is a sufficient test for integrability. In the cases where the criteria for the Painlevé test are met, the analysis may have failed to detect an essential singularity and further analysis would be needed to rigorously prove integrability by:

a) constructing the full set of integrals of the motion,[17] or
b) linearizing the equations, e.g. by the inverse scattering transform,[12] or
c) reducing them to one of Painlevé's transcendental equations.[13,16,18]

This may often be an involved exercise, particularly for the cases where the equation possesses only the "weak" Painlevé property[13,16] (i.e. the leading-order singularity is a movable rational branch point) and a non-obvious set of transformations on both independent and dependent variables may be needed to bring it to a form in which it has the full or "strong" Painlevé property (i.e. the leading-order singularity is a movable pole). In a large number of cases where the equation possesses the full Painlevé property, the associated linear scattering problem or Lax pair has been derived and the resulting Cauchy problem solved by the inverse scattering method. In fact, the usefulness of the Painlevé approach is not limited to integrability prediction, and using the generalized Weiss algorithm[16,19] yields auto-Bäcklund transformations and Lax pairs for the integrable cases. Painlevé analysis also yields a systematic procedure for obtaining special solutions when the equation possesses only the conditional Painlevé property,[20-22] when the compatibility conditions of the Painlevé analysis result in constraint equations for the movable singular manifold which is no longer completely arbitrary.

In this paper, we will consider the Painlevé analysis and the similarity reduced solutions of the Melnikov equations'[23] describing wave interactions in two spatial dimensions:

$$\frac{\partial U}{\partial t} = \frac{\partial U}{\partial y} - 2\kappa \frac{\partial}{\partial x} |\phi|^2 , \tag{1}$$

$$i \frac{\partial \phi}{\partial t} = U\phi + \phi_{xx} . \tag{2}$$

We also consider special analytic solutions of the Zakharov-Kuznetsov equation obtained using Painlevé expansions truncated at various levels.

The remainder of this paper is organized as follows. In Section 2, the Painlevé analysis of Eqs (1) and (2) is carried out, and it is found that these equations possess only the conditional Painlevé property. In Sections 3.1, 3.2, and 3.3 we consider Painlevé expansions truncated at three different levels in an effort to derive singular solutions of Eqs (1) and (2). Stretching transformations[24] and Lie group analysis are employed in Sections 4 and 5 to derive similarity-reduced solutions of Eqs (1) and (2). A Painlevé analysis of these similarity reductions is also considered there. Section 6 considers Painlevé expansions truncated at various levels to obtain and discuss analytic solutions of the conditionally-Painlevé Zakharov-Kuznetsov equation.

## 2 Painlevé analysis

### 2.1 Leading-order and resonance analyses

The behavior of solutions of Eqs (1) and (2) at a movable singular manifold[15]

$$\tilde{\phi}(x, y, t) = 0 \tag{3}$$

is determined by a leading-order analysis, whereby one makes the ansatz[13,16]

$$U \sim a_0 \; \tilde{\phi}^{\alpha}(x, y, t)$$

$$\phi \sim b_0 \; \tilde{\phi}^{\beta}(x, y, t) \tag{4}$$

and balances the most singular or dominant terms. Using (4) in (1) and (2), we find that the only consistent dominant balance is given by

$$\alpha = -2 , \qquad \beta = -1 . \tag{5}$$

In order to simplify further analysis, we use the Kruskal ansatz[13,16]

$$\tilde{\phi}(x, y, t) = y - \psi(x, t) , \tag{6}$$

which is guaranteed by the implicit function theorem whenever the singularity manifold is not a characteristic ($\bar{\phi}_y \neq 0$). Use of (5) and (6) implies that the generalized Laurent expansions about the manifold (3) may be written as:

$$U = \sum_{j=0}^{\infty} a_j(x,t)[y - \psi(x,t)]^{j-2}, \tag{7}$$

$$\phi = \sum_{j=0}^{\infty} b_j(x,t)[y - \psi(x,t)]^{j-1}, \tag{8a}$$

$$\phi^* = \sum_{j=0}^{\infty} b_j^*(x,t)[y - \psi(x,t)]^{j-1}, \tag{8b}$$

where the asterisk denotes complex conjugation. Notice that use of (8) will enable us to transform away the absolute value signs in (1). The resulting equations will then be analytic and Painlevé analysis may be carried out. Also, use of the leading-order terms of (7) and (8) in (1) and (2) yields:

$$a_0 = -2\psi_x^2, \tag{9}$$

$$b_0 b_0^* = \frac{1}{\kappa}(1 + \psi_t)\psi_x. \tag{10}$$

The powers of $[y - \psi(x,t)]$ at which arbitrary coefficients appear in the series (7) and (8), i.e. the so-called resonances, are determined by the "resonance" analysis[13,16] where we set

$$U = a_0(y - \psi)^{-2} + a_r(y - \psi)^{r-2}$$

$$\phi = b_0(y - \psi)^{-1} + b_r(y - \psi)^{r-1}. \tag{11}$$

Using (11) in (1) and (2) and balancing the most singular terms again yields at $0(y - \psi)^{r-3}$:

$$-a_r\{(r-2)\psi_t + (r-2)\} - 2\kappa b_0 b_r^*(r-2)\psi_x - 2\kappa b_0^* b_r \psi_x (r-2) = 0,$$

$$a_r b_0 + a_0 b_r + b_r(r-1)(r-2)\psi_x^2 = 0,$$

and

$$a_r b_0^* + a_0 b_r^* + b_r^*(r-1)(r-2)\psi_x^2 = 0.$$

For $a_r$ and $b_r$ to be arbitrary, the determinant of the matrix of coefficients must vanish, yielding:

$$\begin{vmatrix} (r-2)(1+\psi_t) & 2\kappa b_0^* \psi_x (r-2) & 2\kappa b_0 \psi_x (r-2) \\ b_0 & a_0 + (r-1)(r-2)\psi_x^2 & 0 \\ b_0^* & 0 & a_0 + (r-1)(r-2)\psi_x^2 \end{vmatrix} = 0,$$

which yields the resonances

$$r = -1,\ 0,\ 2,\ 3, \quad \text{and} \quad 4\,. \tag{12}$$

The resonance $r = -1$ corresponds to the arbitrary location of the manifold (3),[13,16] which is the first arbitrary constant in the series (7) and (8). By the Cauchy-Kowaleskaya condition,[25] five arbitrary coefficients or constants are required in order for (7) and (8) to represent valid Laurent expansions for the solutions of (1) and (2) (which comprise two second-order equations coupled to one first-order equation) about (3). The resonances at $r = 2$, 3, and 4 are locations where arbitrary constants may occur and provide the second, third, and fourth arbitrary coefficients in the series (7) and (8). Equation (10) implies that either $b_0$ (or $b_0^*$ may be considered arbitrary, with the other being determined by (10). The resonance at $r = 0$ in (12) corresponds to this arbitrary constant at leading-order. Thus, (7) and (8) have the full complement of arbitrary coefficients and may represent valid generalized Laurent expansions. In order to see whether these expansions are, indeed, single-valued we now proceed to develop these expansions and check the 'resonance conditions'.

### 2.2 Generalized Laurent expansions

We next consider the generalized Laurent expansions (7) and (8) truncated at the highest resonance $r = 4$, i.e. at $j = 4$ in (7) and (8) in the next step of the Painlevé analysis. Using these truncated expansions in (1) and (2), we have a set of equations order-by-order in powers of $\phi$. These equations are given in Appendix A.

Solving (A1), we obtain

$$a_1 = -2\psi_{xx}\,, \tag{13a}$$

$$b_1 = \frac{-b_0\psi_{xx} + ib_0\psi_t - 2b_{0x}\psi_x + 2b_0\psi_{xx}}{-2\psi_x^2}\,. \tag{13b}$$

Using these in (A2a) yields a complicated condition for the function $\psi(x,t)$. Hence, already at this stage, the manifold $\phi(x,y,t) = y - \psi(x,t)$ is not arbitrary. Only special families of manifolds are thus admissible, so that the Melnikov equations (1) and (2) have only the conditional Painlevé property.[13,19]

For cases where the equation possesses only the conditional Painlevé property, truncated Painlevé expansions may be used to derive special analytic solutions.[20-22] We will follow this approach in the next section in attempting to derive special or singular solutions of the Melnikov equations (1) and (2).

## 3 Truncated Painlevé expansions

### 3.1 Truncation at $0(\phi^0)$; Bäcklund transformation

In attempting to derive special solutions of (1) and (2) via the use of truncated Painlevé expansions, we use the truncated expansions

$$U = a_0(x,y,t)[\tilde{\phi}(x,y,t)]^{-2} + a_1(x,y,t)[\tilde{\phi}(x,y,t)]^{-1} + a_2(x,y,t)\,, \tag{14a}$$

$$\phi(x,y,t) = b_0(x,y,t)[\tilde{\phi}(x,y,t)]^{-1} + b_1(x,y,t)\,. \tag{14b}$$

Notice that (6) is not employed here. These expansions are truncated at the constant term in order to obtain auto-Bäcklund transformations[20-22,26] between two sets of solutions of the equations (1) and (2). Using (14) in (1) and (2) yields, after simplification, the equations:

$$\underline{0(\phi^{-3})}: \qquad a_0 = -2\tilde{\phi}_x^2, \tag{15a}$$

$$|b_0|^2 = -\frac{\tilde{\phi}_x}{\kappa}(\tilde{\phi}_y - \tilde{\phi}_t), \tag{15b}$$

$$\underline{0(\phi^{-2})}: \quad a_1 = 2\tilde{\phi}_{xx}, \qquad b_1 = \frac{b_0\tilde{\phi}_{xx} - ib_0\tilde{\phi}_t + 2b_{0x}\tilde{\phi}_x - 2b_0\tilde{\phi}_{xx}}{-2\tilde{\phi}_x^2}, \tag{16ab}$$

$$\underline{0(\phi^{-1})}: \qquad a_{2t} - a_{2y} + 2\kappa\partial_x(b_1 b_1^*) = 0, \qquad a_2(x,y,t)b_1 - ib_{1t} + b_{1xx} = 0. \tag{17ab}$$

Equations (17ab) show that $a_2(x,y,t)$ and $b_1(x,y,t)$ satisfy the same equations (1) and (2) governing $U$ and $\phi$. Thus, Eqs (14) comprise an auto-Bäcklund transformation between two sets of solutions $(U,\phi)$ and $(a_2, b_1)$ of Eqs (1) and (2).

In principle, (15) and (16) may be substituted in (17) yielding two partial differential equations for the singularity manifold $\tilde{\phi}(x,y,t)$ (an overdetermined system). Solving these equations yields the allowed class of $\tilde{\phi}(x,y,t)$ and, via (14), special analytic solutions of (1) and (2). In practice, the resulting equations are involved. Hence, we proceed to the next-order truncation of the series (14).

### 3.2 Truncation at $0(\phi^{-1})$

Setting $a_2 = b_1 = 0$ in (14) and substituting in (1) and (2) yields:

$$\underline{0(\tilde{\phi}^{-3})}: \quad \text{Equations (15)}$$

$$\underline{0(\tilde{\phi}^{-2})}: \quad a_1 = (i\tilde{\phi}_t + \tilde{\phi}_{xx}) + \frac{2b_{0x}^*}{b_0^*}\phi_x$$

$$= \frac{-a_{0t} + a_{0y} - 2\kappa\partial_x(b_0 b_0^*)}{\phi_y - \phi_t}, \tag{18}$$

$$\underline{0(\tilde{\phi}^{-1})}: \quad a_{1t} - a_{1y} = 0, \qquad -ib_{0t} + b_{0xx} = 0. \tag{19ab}$$

In order to facilitate solution of these equations, we define[20] the "dimensions of velocity"

$$C_1 \equiv \frac{\tilde{\phi}_t}{\tilde{\phi}_x}, \qquad C_2 \equiv \frac{\tilde{\phi}_t}{\tilde{\phi}_y},$$

and the parameter

$$V \equiv \frac{\tilde{\phi}_{xx}}{\tilde{\phi}_x}.$$

Notice that for strong Painlevé equations, the equations obtained at various orders in $\tilde{\phi}$ are invariant[19,20,27] under the group of Möbius of homographic transformations $\tilde{\phi} \to \frac{(a\tilde{\phi}+b)}{(c\tilde{\phi}+d)}$. Thus, the use of the invariants of this group, i.e. $C_1$, $C_2$, and the Schwarzian derivative $S \equiv V_x - V^2/2$ enables the algorithmic derivation of the Lax pair[19,20] (the apparently overdetermined system of compatibility equations for the auto-Bäcklund Painlevé expansion truncated at $0(\tilde{\phi}^0)$ collapses into a single equation for strong Painlevé, or presumably integrable, cases; this equation for $\tilde{\phi}$, under a further transformation of variables, yields the Lax pair). For the present conditionally Painlevé equations, the use of $C_1$, $C_2$ and $V$ merely simplifies the equations for the manifold $\tilde{\phi}(x,y,t)$, by giving lower-order equations in terms of these variables.

From (19), we have

$$b_0 = [\tilde{a}e^{lx} + \tilde{b}e^{-lx}]e^{-ilt}, \tag{20a}$$

where $\tilde{a}$, $\tilde{b}$, and $l$ are constants, and

$$a_1 = \alpha_1(x)f(y+t), \tag{20b}$$

with $\alpha_1$ and $f$ being arbitrary functions. Using (20) and (15) in (18) yields, after some manipulations,

$$\tilde{\phi}(x,y,t) = \tilde{\phi}_1(x)\tilde{\phi}_2(y+t),$$

$$a_1 = i\tilde{\phi}_{2t}\tilde{\phi}_1 + \tilde{\phi}_{1xx}\tilde{\phi}_2,$$

$$a_0 = -2\tilde{\phi}_{1x}^2\tilde{\phi}_2^2,$$

$$b_0 = 0.$$

Since $b_0 = 0$, we thus obtain only the trivial solution $\phi = 0$, with

$$U = -\frac{2\tilde{\phi}_{1x}^2}{\tilde{\phi}_1^2} + \left(i\frac{\tilde{\phi}_{2t}}{\tilde{\phi}_2} + \frac{\tilde{\phi}_{1xx}}{\tilde{\phi}_1}\right).$$

### 3.3 Truncation at leading-order

Further truncating (14) by setting $a_1 = 0$ yields equations (15), (19b), and (18) with $a_1 = 0$. Expressing these in terms of $V$, $C_1$, and $C_2$ and solving yields

$$\tilde{\phi}(x,y,t) = \delta(x + C_1 t)$$

$$a_0 = -2\delta^2,$$

$$b_0 = \text{constant}.$$

Note that in order to obtain solutions we have made the assumption[22] that $V$, $C_1$, and $C_2$ are constant. Using these in (1) yields

$$|b_0|^2 = C_1/\kappa$$

and, from (2),

$$-\frac{ib_0 C_1}{(x + C_1 t)^2} = 0.$$

Hence, $b_0 = C_1 = 0$ and we again obtain the trivial solution

$$\phi(x,y,t) = 0.$$

It appears from the above analysis that the Melnikov equations (1) and (2) do not admit simple traveling-wave solutions. This is supported by the fact that, for various other conditionally-Painlevé equations and families of equations,[20-22] the use of truncated Painlevé expansions at various orders (in the same systematic manner as done above) algorithmically yields special analytic solutions of the equations, particularly of various traveling-wave forms. Some of these traveling-wave solutions are of the solitary-wave variety ($U(+\infty) = U(-\infty) = 0$), while others are shock solutions ($U(+\infty) \neq U(-\infty)$). Motivated by the fact that the truncated Painlevé expansions yield only trivial solutions of (1) and (2), we proceed to derive similarity-reduced solutions of (1) and (2) in the hope of finding solutions expressible in terms of elementary functions, elliptic functions, or any of the six Painlevé transcendents,[18] i.e. solutions which do not have movable essential singularities.

## 4 Stretching transformation (dilatation) analysis

In this section, we use the classical method of stretchings[24] to obtain similarity variables and invariant similarity solutions of Eqs (1) and (2). Using the stretched variables

$$\bar{x} = \varepsilon^a x,$$

$$\bar{y} = \varepsilon^b y,$$

$$\bar{t} = \varepsilon^c t$$

$$\bar{U} = \varepsilon^d U, \qquad \bar{\phi} = \varepsilon^e \phi \tag{21}$$

in (1) and (2) yields:

$$\bar{U}_{\bar{t}} - \bar{U}_{\bar{y}} + 2\kappa\partial_{\bar{x}}|\bar{\phi}|^2 = \varepsilon^{d-c}U_t - \varepsilon^{d-b}U_y + 2\kappa\varepsilon^{2e-a}\partial_x|\phi|^2 = 0, \tag{22}$$

$$i\overline{\phi}_{\overline{t}} - \overline{U}\,\overline{\phi} + \overline{\phi}_{\overline{x}\overline{x}} = i\varepsilon^{e-c}\phi_t - \varepsilon^{d+e}U\phi + \varepsilon^{e-2a}\phi_{xx} = 0 \tag{23}$$

For Eqs (1) and (2) to be "constant-conformally invariant" under the transformation (21), i.e. for the transformed operators to be multiples of the original operators, we need

$$b = c\,, \qquad d - b = 2e - a$$

$$2a = c = -d = b\,.$$

Selecting $a = 1$, yields $b = c = 2$, $d = -2$, $e = -3/2$. Hence, from (21),

$$\overline{x} = \varepsilon x\,, \qquad \overline{y} = \varepsilon^2 y\,, \qquad \overline{t} = \varepsilon^2 t$$

$$\overline{U} = \varepsilon^{-2}U\,, \qquad \overline{\phi} = \varepsilon^{-3/2}\phi\,. \tag{24}$$

Clearly,

$$s_1 = \frac{x^2}{y}\,, \qquad s_2 = \frac{t}{y} \tag{25ab}$$

are invariant under transformation (24), i.e. independent of $\varepsilon$. Hence, $s_1$ and $s_2$ are appropriate similarity variables for (1) and (2). Also, by inspection,

$$Uy = \overline{U}\,\overline{y} \quad \text{and} \phi y^{3/4} = \overline{\phi}\,\overline{y}^{3/4}$$

are invariant under (24), and hence are functions of $s_1$ and $s_2$. Thus, the self-similar solution of (1) and (2) obtained via the method of stretchings is

$$U(x,y,t) = y^{-1}f(s_1,s_2)\,, \tag{26}$$

$$\phi(x,y,t) = y^{-3/4}g(s_1,s_2)\,. \tag{27}$$

This solution could also have been obtained using dimensional analysis. For self-similar solutions of the form (26)-(27), equations (1) and (2) yields two similarity reduced coupled partial differential equations:

$$\frac{\partial f}{\partial s_2} + f + s_1\frac{\partial f}{\partial s_1} + s_2\frac{\partial f}{\partial s_2} + 4\kappa\sqrt{s_1}\left(g^{*}\frac{\partial g}{\partial s_1} + g\frac{\partial g^{*}}{\partial s_1}\right) = 0\,,$$

$$i\frac{\partial g}{\partial s_2} = fg + 2\frac{\partial g}{\partial s_1} + 4z\frac{\partial^2 g}{\partial s_1^2}\,. \tag{28}$$

Further invoking Lie point symmetry of (28) yields that

$$f = s_1^{-1}F(T)\,, \qquad g = s_1^{-3/4}G(T)\,, \tag{29}$$

where

$$T \equiv s_2/s_1 = t/x^2\,. \tag{30}$$

The functions $F(T)$ and $G(T)$ satisfy the coupled ordinary differential equations

$$\frac{dF}{dT} - 6\kappa GG^* - 4\kappa T \partial_T (GG^*) = 0\,, \tag{31a}$$

$$4i\frac{dG}{dT} = 4FG + 15G + 48T\frac{dG}{dT} + 16T^2\frac{d^2G}{dT^2}\,, \tag{31b}$$

$$-4i\frac{dG^*}{dT} = 4FG^* + 15G^* + 48T\frac{dG^*}{dT} + 16T^2\frac{d^2G^*}{dT^2}\,. \tag{31c}$$

The similarity-reduced solutions of (1) and (2) given in (26)-(27) and (29) are in terms of functions $F(T)$ and $G(T)$ satisfying Eqs (31). In order to determine the nature of these functions, we perform a Painlevé analysis of Eqs (31).

Expanding $F(T)$ and $G(T)$ as

$$F(T) = \sum_{n=0}^{\infty} a_n (T - T_0)^{n-2}\,, \qquad G(T) = \sum_{n=0}^{\infty} b_n (T - T_0)^{n-1}\,, \tag{32ab}$$

where the leading-order analysis has already been incorporated and the singularity manifold $T - T_0$ is arbitrary, the leading-order coefficients are

$$a_0 = -8T_0^2\,, \qquad |b_0|^2 = -\frac{2T_0}{\kappa}\,. \tag{33ab}$$

Proceeding to the resonance analysis of (31), the resonance polynomial at $0(T - T_0)^{r-3}$ yields the resonances

$$r = -1,\ 0,\ 2,\ 3,\ 4$$

as for the original Melnikov equations (1) and (2). As discussed earlier for (1) and (2), the resonances $r = -1, 2, 3, 4$ correspond respectively to the arbitrary location $T_0$ of the singularity manifold, and to three arbitrary coefficients in the Laurent expansions (32). Also, the fifth resonance $r = 0$ corresponds to leading-order and, via (33), a fifth arbitrary coefficient occurs (either $b_0$ or $b_0^*$). Using (32) and (33) in (31), we solve for $a_1$, $b_1$ at next-to-leading order. At the next order in $(T - T_0)$, a constraint condition on $(T - T_0)$ results. Thus Eqs (31) have only the conditional Painlevé property, and are not strong Painlevé. Hence, the similarity solutions (26)-(27) and (29) of (1) and (2) may not be expressed in terms of elliptic functions or any of the six Painlevé transcendents by transformations of (31). These solutions will, in general have movable essential singularities.

## 5 Lie group analysis

In this section, we continue our analysis of similarity solutions by using Lie group analysis to obtain similarity-reduced solutions of (1) and (2). We follow the treatments in Bluman and Kumei[10] and Bluman and Cole.[2] Alternatively, the differential forms approach of Harrison and Estabrook[28] may be used. Consider a one-parameter ($\varepsilon$) group of infinitesimal transformations of $(x, y, t, U, \phi)$:

$$U^* = U + \varepsilon\eta(x,y,t,U,\phi), \tag{34a}$$
$$\phi^* = \phi + \varepsilon\gamma(x,y,t,U,\phi), \tag{34b}$$
$$x^* = x + \varepsilon\zeta(x,y,t,U,\phi), \tag{34c}$$
$$y^* = y + \varepsilon\rho(x,y,t,U,\phi), \tag{34d}$$
$$t^* = t + \varepsilon\tau(x,y,t,U,\phi), \tag{34e}$$
$$U^*_{t^*} = U_t + \varepsilon\eta^t, \tag{34f}$$
$$U^*_{x^*} = U_x + \varepsilon\eta^x, \tag{34g}$$
$$U^*_{y^*} = U_y + \varepsilon\eta^y, \tag{34h}$$
$$\phi^*_{t^*} = \phi_t + \varepsilon\gamma^t, \tag{34i}$$
$$\phi^*_{x^*} = \phi_x + \varepsilon\gamma^x, \tag{34j}$$
$$\phi^*_{x^*x^*} = \phi_{xx} + \varepsilon\gamma^{xx}. \tag{34k}$$

The extensions $\eta^t$, $\eta^x$, $\eta^y$, $\gamma^t$, $\gamma^x$, and $\gamma^{xx}$ may be computed using the results in Ref. 2 for two dependent and three independent variables. These extensions are given in Appendix B.

Equations (1) and (2) are invariant under the Lie group transformation (34) if $U^*(x^*,y^*,t^*)$ and $\phi^*(x^*,y^*,t^*)$ satisfy the same equations (1) and (2) as $U(x,y,t)$ and $\phi(x,y,t)$. Substituting (34) and (1) and (2), yields to first order in $\varepsilon$:

$$\eta^t - \eta^y + 2\kappa\left[\phi_.\gamma^x + \gamma_.\phi_x + \phi(\gamma^x)^* + \gamma\phi^*_x\right] = 0, \tag{35a}$$

$$i\gamma^t = U\gamma + \eta\phi + \gamma^{xx}. \tag{35b}$$

The infinitesimals $\eta$, $\gamma$, $\zeta$, $\rho$, and $\tau$ are computed by using the extensions $\eta^t$, $\eta^x$, $\eta^y$, $\gamma^t$, $\gamma^x$, and $\gamma^{xx}$, collecting coefficients of like derivative terms of $U$ and $\phi$ in (35) and equating them to zero. Solving these "determining equations" in the standard way (note that, from the forms of (1) and (2), $\frac{\partial\zeta}{\partial U} = \frac{\partial\tau}{\partial U} = 0$ using the theorems in Ref. 2) yields:

$$\eta = 0, \tag{36a}$$

$$\gamma = 0, \tag{36b}$$

Clearly, Eqs (36) imply that $U = U^* =$ constant and $\phi = \phi^* =$ constant. Also, one of $U$ and $\phi$ must be zero. Thus, the Lie group analysis yields only trivial solutions of the Melnikov equations (1) and (2).

Since the classical Lie group technique yields only trivial similarity reductions of the Melnikov equations, the similarity reductions of Sec. IV may be recaptured, for instance, by using the direct method of Ref. 11. This method has also been shown[29] to be linked to the nonclassical method of Ref. 6 and to 'conditional symmetries' which leave only some subsets of solutions invariant.

# 6 Auto-Bäcklund transformation and special solutions of the Zakharov-Kuznetsov equation

Nonlinear evolution of ion-acoustic waves in a magnetized plasma is governed[30-33] by the Zakharov-Kuznetsov (ZK) equation

$$U_t + UU_x + U_{xxx} + U_{xyy} = 0\,. \tag{37}$$

Zakharov and Kuznetsov[30] demonstrated balancing of the nonlinearity by dispersive effects, while Laedke and Spatschek[31] developed a systematic perturbation theory. A Lagrangian formulation of (37) was given in 32, while Ref. 33 considered various analytical properties of (37) such as solitary wave solutions, conservation laws and recurrence of solutions.

The ZK equation possesses solitary wave solutions[33]

$$U = 3\mu\, sech^2\left\{\frac{1}{2}\mu^{1/2}[1+\nu^2]^{-1/2}\zeta\right\}, \tag{38}$$

with

$$\zeta \equiv x + \nu y - \mu t\,. \tag{39}$$

The constants $\nu$ and $\mu$ are arbitrary and provide a two-parameter family of solutions and may be chosen to fit physical data. Although the solitary-wave solutions (38) have the hyperbolic secant form typical of 1- and N-soliton solutions of completely integrable equations such as the KdV equation,[34] there is strong evidence that they are only solitary waves and not solitons. There is evidence[31,32] suggesting that (37) is non-integrable, with solitary waves interacting inelastically. Also, (37) possesses only three distinct polynomial conservation laws,[33] instead of the infinite number typical of integrable equations.

## 6.1 Auto-Bäcklund transformation

For the ZK equation, the leading order behavior about a movable singular manifold

$$\tilde{\phi}(x,t) = 0 \tag{40a}$$

has the form

$$U(x,t) = U_0(x,t)[\tilde{\phi}(x,t)]^{-2} \tag{40b}$$

(from the dominant balance between the $UU_x$, $U_{xyy}$ and $U_{xxx}$ terms). Substituting the Painlevé expansion

$$U(x,t) = U_0(x,t)[\tilde{\phi}(x,t)]^{-2} + U_1(x,t)[\tilde{\phi}(x,t)]^{-1} + U_2(x,t) \tag{41}$$

truncated at the constant or $\phi^0$ term in (37) yields a set of coupled nonlinear partial differential equations order by order in terms of $\phi$. These equations are given in Appendix C. Equation (C6) shows that $U_2$ satisfies (37), so that (41) (with $U_0$ given by (42) below) is an auto-Bäcklund transformation between two solutions $U(x,t)$ and $U_2(x,t)$ of (37). Solving equations (C1) and (C2):

$$U_0 = -12\phi_x^2(1 + C_3^2)\,, \tag{42}$$

$$U_1 = 12V\phi_x(1 + C_3^2)\,, \tag{43}$$

Here, we have introduced[20]

$$\tilde{V} \equiv \frac{\tilde{\phi}_{yy}}{\tilde{\phi}_x}\,. \tag{44}$$

and the additional dimension of velocity parameter

$$C_3 = \frac{C_1}{C_2}\,. \tag{45}$$

The general solutions of (C2)-(C6) expressed in terms of $V$, $\tilde{V}$ and $C_i$ subject to the compatibility condition ($\tilde{\phi}_{xxt} = \tilde{\phi}_{txx}$)

$$V_t = C_{1xx} + C_{1x}V + C_1V_x \tag{46}$$

is virtually impossible (this involves nonlinear equations for $V$ and $C_i$). Hence, we make the assumption that $V$, $\tilde{V}$ and the $C_i$ are constants in order to obtain equations that may be solved in order to obtain special analytical solutions of (37). The condition (46) is then automatically satisfied. This assumption has been used to obtain (43) from (C2). Solving (C3) under this assumption yields

$$U_2 = -(C_1 + V^2 + C_3^2V^2)\,, \tag{47}$$

while (C4)-(C6) are identically satisfied. Using (42), (43) and (47) in (41) yields the special analytical solutions

$$U = -C_1 - V^2 - C_3^2V^2 - \frac{12c^2e^{2V(C_1t + x + C_3y)}}{\left[ce^{V(C_1t + x + C_3y)} + c_4\right]^2} + \frac{12ce^{V(C_1t + x + C_3y)}(1 + C_3^2)V^2}{ce^{V(C_1t + x + C_3y)} + c_4} \tag{48}$$

of the ZK equation (37). Note that $C_1, C_3$ and $V$ are arbitrary constants. The solutions (48) represent a new three-parameter family of bidirectional (depending on the sign of $C_1C_3$) traveling-wave solutions (in the $x$ and $y$ directions) of (37), parameterized by $C_1, C_3$, and $V$, not obtainable by any obvious traveling-wave reduction of (37) to an ordinary differential equation. For different choices of the constants $V$, $C_1, C_3$, one may fit different sets of numerical or physical data. The traveling-wave solutions (48) are a distinct family of solutions from those in (38). In order to explore the nature of the new solutions (48), consider the case $V > 0$ (say). For this case, $U(x \to \infty) \to$ constant, and $U(x \to -\infty) \to 0$. Thus, $U(x \to \infty) \neq U(x \to -\infty)$ and the solutions (48) correspond to shock solutions,[35] unlike the solutions (38) which are solitary wave solutions.

One may use the solutions (48) (or the solutions (38)) as the solution $U_2$ of Eq. (37) in the auto-Bäcklund transformation (41). Re-solving (C3)-(C6) for a new class of manifolds $\tilde{\phi}(x,t)$ in the next step of the auto-Bäcklund iteration would, in principle, yield new solutions of Eq. (37) via (48). As one would expect from the earlier discussion, the equations resulting from (C3)-(C6) when this procedure is followed are too complicated to solve in full generality. Assuming $C_i$ and $V$ constant yields $V, C_i = 0$ and thus, $U_0 = U_1 = 0$ from (42) and (43). Hence, from (41), $U = U_2$ and we obtain only the class of solutions (48) (or (38)) that we started with. Such behavior has also been observed earlier in the context of other equations.[20-22]

### 6.2 Next-to-leading-order truncation

Truncating (41) further in the hope of obtaining other analytic solutions, we substitute the next-to-leading-order truncation

$$U(x,t) = \frac{U_0(x,t)}{[\phi(x,t)]^2} + \frac{U_1(x,t)}{\phi(x,t)} \tag{49}$$

in (37). The resulting equations are given by (C1)-(C5) with $U_2 = 0$ (Eqn (C6) is automatically satisfied). Using (42) for $U_0$, and (43) for $U_1$ (these remain unchanged), (C3)-(C5) yield $24(1+C_3^2)(C_1+V^2+C_3^2V^2)\phi_x^3 = 0$, $36(1+C_3^2)(C_1+V^2+C_3^2V^2)V\phi_x^2 = 0$, and $12(1+C_3^2)(C_1+V^2+C_3^2V^2)V^2\phi_x = 0$ respectively. Clearly, either (a) $C_3 = \pm i$, or (b) $C_3, V$ arbitrary and $C_1 = -V^2(1+C_3^2)$. In case (b), $\phi(x,y,t) = ce^{V(x+C_3y-V^2(1+C_3^2)t)} + c_4$. Hence, using (42), (43) in (49) yields the special solutions

$$U = \frac{12cc_4(1+C_3^2)V^2e^{V[V^2(1+C_3^2)t+x+C_3y]}}{\left[ce^{V(x+C_3y)} + C_2e^{V^3t(1+C_3^2)}\right]^2} \tag{50}$$

of (37). Note that $C_3$ and $V$ are arbitrary constants, so that (50) represents another new two-parameter family of waves moving rightward along $x$ with phase-speed $V^2(1+C_3^2)$ and moving towards positive $y$ at phase-speed $V^2(1+C_3^2)/C_3$ for $C_3 > 0$ (and to negative $y$ at speed $-V^2(1+C_3^2)/C_3$ for $C_3 < 0$). Once again, $C_3$ and $V$ may be chosen to fit various sets of numerical or experimental data. As for (48), solutions (50) correspond to shock solutions[35] with $U(\infty) \neq U(-\infty)$ unlike the solitary wave solutions (38). However, (48) and (50) are distinct families of analytical solutions. In particular, (48) is a three-parameter $(C_1, C_3, V)$ family of solutions, while (50) is a two-parameter $(C_3, V)$ family of solutions of (37). For example, the phase-speed in the $x$-direction for the solutions (48) is $C_1$ which may be along either the positive or negative directions, while the solutions (50) always represent waves moving rightward along $x$ with speed $V^2(1+C_3^2)$. Similar differences in the nature of the $y$-propagation exist between the solutions (48) and (50).

For the case (a) above where $C_3 = -i$, one has $\tilde{\phi}_y = \pm i\tilde{\phi}_x$ or $\tilde{\phi}(x,y,t) = \tilde{\phi}(x \pm iy)f(t)$. Thus, (37) admits general solutions of this same functional form. For instance, one may consider $\phi(x,y,t) = ce^{V(x \pm iy + C_1t)} + C_2$ and obtain additional special analytical solutions of

(37) on using this, with (42) and (43), in (49). These may be considered in analogy with the earlier treatments of (48) and (50) and we do not consider this further. Notice that for real or complex $C_1$ these solutions will correspond to growing or damping solutions. For instance, with $C_1 \equiv C_r + iC_i$, the solutions will have the form $e^{Vx}e^{i(\pm y + C_i t)}e^{C_r t}$. In terms of propagation along $y$, these solutions will grow or damp depending on whether $C_r > 0$ or $C_r < 0$.

### 6.3 Leading-order truncation

Truncating (41) at leading-order in looking for additional solutions of (37), i.e. setting $U_1 = U_2 = 0$ in (41) and (C1)-(C4) ((C5) and (C6) are automatically satisfied), one obtains (from (C2) through (C4))

$$C_3 = \pm i \tag{51}$$

again. Thus $\phi(x,y,t) = \phi(x \pm iy)f(t)$ and using this and (42) (obtained from (C1)) in (41) (with $U_1 = U_2 = 0$) yields the general class of solutions

$$U(x,y,t) = \frac{\phi(x \pm iy)}{\phi(t) + k} \tag{52}$$

of (37). This class of solutions may be obtained much more readily by considering traveling-wave reductions of (37). Depending on the specific function $\phi$ used, (52) may represent oscillatory or traveling-wave solutions of (37). For instance, if $\phi(z) = e^z$, (52) yields $U(x,y,t) = e^x(\cos y \pm i \sin y)e^{-t}$, corresponding to traveling waves along $x$ and damped oscillations in $y$.

## Appendix A

The equations obtained using (7) and (8) truncated at $j = 4$ in (1) and (2), yield at various orders in $\phi$:

$$\underline{0(\phi^{-2})}: \quad a_1(1 + \psi_t) + (b_1.2\kappa b_0^* \phi_x + c.c.) = -a_{0t} - 2\kappa\partial_x(b_0 b_0^*), \tag{A1a}$$

$$a_1 b_0 + a_0 b_1 - i b_0 \psi_t + 2 b_{0x}\psi_x + b_0 \psi_{xx} = 0, \tag{A1b}$$

$$\underline{0(\phi^{-1})}: \quad a_{1t} = 2\kappa\left[(b_1 b_0^*)_x + (b_0 b_1^*)_x\right], \tag{A2a}$$

$$a_0 b_2 + a_2 b_0 = i b_{0t} - b_{0xx} - a_1 b_1, \tag{A2b}$$

$$\begin{aligned} \underline{0(\phi^{0})}: \quad & -a_3(1 + \psi_t) - 2\kappa b_3 b_0^* \psi_x - 2\kappa b_0 b_3^* \psi_x \\ & = -a_{2t} - 2\kappa b_{2x} b_0^* - 2\kappa b_{1x} b_{1x}^* \\ & -2\kappa b_{0x} b_2^* - 2\kappa b_2 b_{0x}^* - 2\kappa b_1 b_{1x}^* \\ & -2\kappa b_0 b_{2x}^* + 2\kappa b_2 b_1^* \psi_x - 2\kappa b_1 b_2^* \psi_x, \end{aligned} \tag{A3a}$$

$$a_0 b_3 = i b_{1t} - i b_2 \psi_t - b_{1xx} + 2 b_{2x} \psi_x$$
$$+ b_2 \psi_{xx} - a_2 b_1 - b_2 a_1 , \qquad (A3b)$$

$$\underline{0(\phi)}: \quad 2a_4(1+\psi_t) - b_{4}.(4\kappa b_0^* \psi_x) - b_4^*(4\kappa b_0 \psi_x)$$
$$= -a_{3t} - 2\kappa b_{3x} b_0^* - 2\kappa b_{2x} b_1^* - 2\kappa b_{1x} b_2^* - 2\kappa b_{0x} b_3^*$$
$$- 2\kappa b_3 b_{0x}^* - 2\kappa b_2 b_{1x}^* - 2\kappa b_1 b_{2x}^* - 2\kappa b_0 b_{3x}^*$$
$$+ 4\kappa b_1^* b_3 \psi_x + 2\kappa b_2 b_2^* \psi_x + 2\kappa b_2 b_2^* \psi_x + 4\kappa b_1 b_3^* \psi_x , \qquad (A4a)$$

$$a_4 b_0 + b_4 . b \psi_x^2 = i b_{2t} - 2 i b_3 \psi_t - b_{2xx} + 4 b_{3x} \psi_x$$
$$- 2 b_3 \psi_{xx} - a_3 b_1 - a_2 b_2 - a_1 b_3 . \qquad (A4b)$$

## Appendix B

The extensions $\eta^x, \eta^t, \eta^y, \gamma^x, \gamma^t$, and $\gamma^{xx}$ are:

$$\eta^x = \frac{\partial\eta}{\partial x} + \left(\frac{\partial\eta}{\partial U}\cdot\frac{\partial U}{\partial x} + \frac{\partial\eta}{\partial\phi}\cdot\frac{\partial\phi}{\partial x}\right) - \left(\frac{\partial U}{\partial x}\cdot\frac{\partial\zeta}{\partial x} + \frac{\partial U}{\partial y}\cdot\frac{\partial\rho}{\partial x} + \frac{\partial U}{\partial t}\cdot\frac{\partial\tau}{\partial x}\right)$$
$$- \left[\frac{\partial U}{\partial x}\cdot\left(\frac{\partial\zeta}{\partial U}\cdot\frac{\partial U}{\partial x} + \frac{\partial\zeta}{\partial\phi}\cdot\frac{\partial\phi}{\partial x}\right) + \frac{\partial U}{\partial y}\left(\frac{\partial\rho}{\partial U}\cdot\frac{\partial U}{\partial x} + \frac{\partial\rho}{\partial\phi}\cdot\frac{\partial\phi}{\partial x}\right)\right.$$
$$\left. + \frac{\partial U}{\partial t}\left(\frac{\partial\tau}{\partial U}\cdot\frac{\partial U}{\partial x} + \frac{\partial\tau}{\partial\phi}\cdot\frac{\partial\phi}{\partial x}\right)\right] \qquad (B1)$$

Similarly,

$$\eta^t = \eta_t + (\eta_U . U_t + \eta_\phi \phi_t) - (\zeta_t U_x + U_y \rho_t + U_t . \tau_t)$$
$$- [U_x(\zeta_U . U_t + \zeta_\phi \phi_t) + U_y(\rho_U . U_t + \rho_\phi \phi_t) + U_t(\tau_U U_t + \tau_\phi \phi_t)] \qquad (B2)$$

$$\eta^y = \eta_y + (\eta_U . U_y + \eta_\phi \phi_y) - (U_x . \zeta_y + U_y \rho_y + U_t \tau_y)$$
$$- [U_x(\zeta_U . U_y + \zeta_\phi \phi_y) + U_y(\rho_U . U_y + \rho_\phi \phi_y) + U_t(\tau_U U_y + \tau_\phi \phi_y)] \qquad (B3)$$

$$\gamma^x = \gamma_x + (\gamma_U . U_x + \gamma_\phi \phi_x) - \left(\frac{\partial\phi}{\partial x}\cdot\zeta_x + \phi_y . \rho_x + \phi_t . \tau_x\right)$$
$$- [\phi_x(\zeta_U U_x + \zeta_\phi \phi_x) + \phi_y(\rho_U U_x + \rho_\phi \phi_x) + \phi_t(\tau_U U_x + \tau_\phi \phi_x)] \qquad (B4)$$

$$\gamma^t = \gamma_t + (\gamma_U U_t + \gamma_\phi \phi_t) - (\phi_x . \zeta_t + \phi_y \rho_t + \phi_t \tau_t)$$
$$- [\phi_x(\zeta_U U_t + \zeta_\phi \phi_t) + \phi_y(\rho_U U_t + \rho_\phi \phi_t) + \phi_t(\tau_U U_t + \tau_\phi \phi_t)] \qquad (B5)$$

$$\gamma^{xx} = \gamma_{xx} + (\gamma_{xU} . U_x + \gamma_{x\phi}\phi_x) + (\gamma_{xU} U_x + \gamma_{x\phi}\phi_x) - (\phi_x \zeta_{xx} + \phi_y \rho_{xx} + \phi_t \tau_{xx})$$
$$+ (U_{xx} . \gamma_U + \phi_{xx} . \gamma_\phi) - (\phi_{xx}\zeta_x + \phi_{xy}\rho_x + \phi_{xt}\tau_x) - (\phi_{xx} . \zeta_x + \phi_{xy} . \rho_x + \phi_{xt}\tau_x)$$
$$+ [U_x U_x \gamma_{UU} + U_x \phi_x . \gamma_{U\phi} + \phi_x U_x \gamma_{U\phi} + \phi_x^2 \gamma_{\phi\phi}] - [2\phi_x(\zeta_{xU} U_x + \zeta_{x\phi}\phi_x)$$
$$+ 2\phi_y(\rho_{xU} U_x + \rho_{x\phi}\phi_x) + 2\phi_t(\tau_{xU} U_x + \tau_{x\phi}\phi_x)] - U_{xx}[\zeta_U . \phi_x + \rho_U \phi_y + \tau_U . \phi_t]$$
$$- \phi_{xx}[\zeta_\phi \phi_x + \rho_\phi \phi_y + \tau_\phi \phi_t] - 2\phi_{xx}[\zeta_U . U_x + \zeta_\phi \phi_x] - 2\phi_{xy}[\rho_U U_x + \rho_\phi \phi_x]$$
$$- 2\phi_{xt}[\tau_U U_x + \tau_\phi \phi_x] - \phi_x[\zeta_{UU} U_x U_x + \zeta_{U\phi} U_x \phi_x + \zeta_{\phi U}\phi_x U_x + \zeta_{\phi\phi}\phi_x \phi_x]$$
$$- \phi_y[\rho_{UU} U_x U_x + \rho_{U\phi} U_x \phi_x + \rho_{\phi U}\phi_x U_x + \rho_{\phi\phi}\phi_x \phi_x] - \phi_t[\tau_{UU} U_x^2 + 2\tau_{U\phi} U_x \phi_x + \tau_{\phi\phi}\phi_x^2] \quad (B6)$$

## Appendix C

The equations at various orders in $\phi$ resulting from using (41) in (37) are:

$$\underline{0(\phi^{-5})}: \quad -2U_0^2\phi_x - 24U_0\phi_y^2\phi_x - 24U_0\phi_x^3 = 0\,, \qquad (C1)$$

$$\underline{0(\phi^{-4})}: \quad -3U_0 U_1 \phi_x - 6U_1\phi_y^2\phi_x + 12\phi_y U_{0y}\phi_x + 6U_0\phi_{yy}\phi_x$$
$$- 6U_1\phi_x^3 + U_0 U_{0x} + 6\phi_y^2 U_{0x} + 18\phi_x^2 U_{0x}$$
$$+ 12U_0\phi_y\phi_{xy} + 18U_0\phi_x\phi_{xx} = 0\,, \qquad (C2)$$

$$\underline{0(\phi^{-3})}: \quad -2U_0\phi_t - U_1^2\phi_x - 2U_0 U_2\phi_x + 4\phi_y U_{1y}\phi_x$$
$$+ 2U_1\phi_{yy}\phi_x - 2U_{0yy}\phi_x + U_1 U_{0x} - 2\phi_{yy}U_{0x} + U_0 U_{1x}$$
$$2\phi_y^2 U_{1x} + 6\phi_x^2 U_{1x} + 4U_1\phi_y\phi_{xy} - 4U_{0y}\phi_{xy}$$
$$- 4\phi_y U_{0xy} - 2U_0\phi_{xyy} + 6U_1\phi_x\phi_{xx} - 6U_{0x}\phi_{xx}$$
$$- 6\phi_x U_{0xx} - 2U_0\phi_{xxx} = 0\,, \qquad (C3)$$

$$\underline{0(\phi^{-2})}: \quad -U_1\phi_t + U_{0t} - U_1U_2\phi_x - U_{1yy}\phi_x + U_2U_{0x}$$
$$+ U_1U_{1x} - \phi_{yy}U_{1x} + U_0U_{2x} - 2U_{1y}\phi_{xy} - 2\phi_y U_{1xy}$$
$$- U_1\phi_{xyy} + U_{0xyy} - 3U_{1x}\phi_{xx} - 3\phi_x U_{1xx}$$
$$- U_1\phi_{xxx} + U_{0xxx} = 0\,, \qquad (C4)$$

$$\underline{0(\phi^{-1})}: \quad U_{1t} + U_2U_{1x} + U_1U_{2x} + U_{1xyy} + U_{1xxx} = 0\,, \qquad (C5)$$

$$\underline{0(\phi^{0})}: \quad U_{2t} + U_2U_{2x} + U_{2xyy} + U_{2xxx} = 0\,. \qquad (C6)$$

## References


1. Lie, S., Vorlesungen über Differentialgleichungen mit Bekannten Infinitesimalen Transformationen. Teuber, Leipzig (1891) (reprinted by Chelsea, New York (1967)).

2. Bluman, G. W. and Cole, J. D., Similarity Methods for Differential Equations, Springer-Verlag (Applied Mathematical Sciences), Berlin (1974); Bluman, G. W. and Kumei, S., Symmetries and Differential Equations (Springer, New York, 1989).

3. Olver, P. J., Applications of Lie Groups to Differential Equations, Springer-Verlag (Graduate Texts in Mathematics), New York (1986).

4. Winternitz, P., Nonlinear Phenomena, Springer-Verlag (Lecture Notes in Physics, v. 189), Berlin (1983).

5. Schwarz, F., SIAM Review 30, 450-481 (1988).

6. Bluman, G. W. and Cole, J. D., J. Math Mech. 18, 1025-42 (1969).

7. Olver, P. J. and Rosenau, P., Physics Letter 114A, 107-112 (1986); and in SIAM J. Appl. Math. 47, 263-78 (1987).

8. Noether, E., Nachr. König. Gesell. Wissen Göttingen, Math. Phys. K1, 235 (1918).

9. Anderson, R. L. and Ibragimov, N. H., Lie-Bäcklund Transformations in Applications, SIAM, Philadelphia (1979).

10. Bluman, G. W., Kumei, S. and Reid, G. J., J. Math. Phys. 29, 806-11 (1988).

11. Clarkson, P. A. and Kruskal, M. D., J. Math. Phys. 30, 2201-2213 (1989); Clarkson, P. A., J. Phys. A: Math Gen. 22, 3821-3848 (1989); 22, 2355-2367 (1989).

12. Clarkson, P. A. and Winternitz, P., Physica D49, 257-272 (1991).

13. Ramani, A., Grammaticos, B. and Bountis, T., Phys. Reports 180, 160-244 (1989).

14. Ablowitz, M. J., Ramani, A. and Segur, H., J. Math. Phys. 21, 715-721, 1006-1015 (1980).

15. Weiss, J., Tabor, M. and Carnevale, G., J. Math. Phys. 24, 522-526 (1983).

16. Tabor, M., Chaos and Integrability in Nonlinear Dynamics, Wiley, New York (1989).

17. Hietarinta, J., Phys. Reports 147, 87-154 (1987).

18. Ince, E. L., Ordinary Differential Equations, Dover, New York (1956).

19. Weiss, J., J. Math. Phys. 24, 1405-1413 (1983); 25, 13-24, 2226-2235 (1984).

20. Conte, R. and Musette, M., J. Phys. A: Math. and Phys. 22, 169-177 (1989); Kudryashov, N. A., Phys. Letters 147A, 287-291 (1990); Conte, R., Phys. Lett. A140, 383 (1989); Musette, M. and Conte, R., J. Math. Phys. 32, 1450 (1991).

21. Cariello, F. and Tabor, M., Physica D39, 77-94 (1989).

22. Choudhury, S. Roy, J. Nonlinear Analysis, 18, 445-459 (1992); and in Can. J. Phys. 70, 273-281, 595-602 (1992).

23. Melnikov, V. K., Math. Sbornik 121, 469-498 (1983); and in Lett. Math. Phys. 7, 129 (1983); and in J. Math. Phys. 28, 2603 (1987); and in Comm. Math. Phys. 112, 639 (1987).

24. Logan, J. D., Applied Mathematics (Wiley-Interscience, New York, 1987).

25. Hille, E., Analytic Function Theory (Ginn, Boston, 1962).

26. Steeb, W. H. and Euler, N., Nonlinear Evolution Equations and the Painlevé Test (World Scientific, Singapore, 1988).

27. Musette, M. and Conte, R., J. Math. Phys. 32, 1450 (1991); Conte, R., Phys. Lett. A140, 383 (1989).

28. Harrison, B. K. and Estabrook, F. B., J. Math. Phys. 12, 653 (1971).

29. Levi, D. and Winternitz, P., J. Phys. A22, 2915 (1989).

30. V. E. Zakharov and E. M. Kuznetsov, Sov. Phys. JETP 39, 285-286 (1974).

31. E. W. Laedke and K. H. Spatschek, Phys. Fluids 25 (1982), 985; and in J. Plasma Phys. 28, 469-474 (1982).

32. E. Infeld, J. Plasma Phys. 33, 171-177 (1985).

33. B. K. Shivamoggi, J. Plasma Phys. 41, 83-88 (1989).

34. Ablowitz, M. J. and Segur, H., 1981, Solitons and the Inverse Scattering Transform (Philadelphia: Society for Industrial and Applied Mathematics); Dodd, R. K., Eilbeck, J. C., Gibbon, J. D. and Morris, H. C., 1982 Solitons and Nonlinear Wave Equations (London: Academic).

35. Chang, H. C., Phys. Fluids 29, 3142-3146 (1986).

# Chapter 5

# Second-order loading on a pair of vertical cylinders

**M. Rahman, D.D. Bhatta**

*Department of Applied Mathematics, Technical University of Nova Scotia, PO Box 1000, Halifax, Nova Scotia, Canada B3J 2X4*

## Abstract

An estimation of the hydrodynamic forces to the second order is presented here. Special attention is paid to derive the quadratic forces for a pair of bottom mounted, surface piercing circular cylinders in waves of arbitrary uniform depth. The theoretical results are based on the large spacing approximation and the method involves replacing scattered diverging waves by plane waves. The theoretical results are presented to illustrate the influence of the various wave and structural parameters on the hydrodynamic forces on each of the cylinders. The amplification factors for the first and second order drift forces are computed and presented in tabular form. The second order forces on a single cylinder are also computed numerically and presented in graphical forms for various wave parameters.

## Keywords

Diffraction, radiation, velocity potential, hydrodynamic loading, first order, second order, pair of cylinders

## List of symbols

| | |
|---|---|
| $\Phi$ | velocity potential (real) |
| $\phi$ | velocity potential (complex) |
| $\psi$ | potential due to radiation |
| $\phi_I^{(1)}$ | linear incident velocity potential |
| $\phi_S^{(1)}$ | linear scattered velocity potential |
| $\phi_I^{(2)}$ | second order incident velocity potential |
| $\phi_S^{(2)}$ | second order scattered velocity potential |
| $\sigma$ | frequency |

| | |
|---|---|
| $k$ | wavenumber |
| $h$ | water depth |
| $g$ | acceleration due to gravity |
| $A$ | incident wave amplitude |
| $x, y, z$ | cartesian coordinates |
| $r, \theta, z$ | cylindrical polar coordinates |
| $J_n$ | Bessel function of order $n$ |
| $H_n^{(2)}$ | Modified Bessel function of order $n$ |
| $F_x^{(1)}$ | first order force in $x$-direction |
| $F_y^{(1)}$ | first order force in $y$-direction |
| $F_q^{(2)}$ | quadratic force on single cylinder |
| $F_{wx}^{(2)}$ | water line force in $x$-direction |
| $F_{wy}^{(2)}$ | water line force in $y$-direction |
| $F_{dx}^{(2)}$ | dynamic force in $x$-direction |
| $F_{dy}^{(2)}$ | dynamic force in $y$-direction |

# 1 Introduction

The problem of hydrodynamic interactions within a group of large multi-component offshore structures subjected to ocean waves has been of much interest among scientists and engineers for many years. From a practical consideration it is extremely important to be able to predict the wave loads on multicolumn structures in waves. When many offshore structures are placed together in a configuration, the hydrodynamic loads on individual components may be significantly different from the loading they each would experience in isolation. This is simply because of the hydrodynamic coupling between the bodies. Ohkusu (1973) used the method of multiple scattering in which the scattered potential is determined from each scattering event within the cylinder group. Spring and Monkmeyer (1974) adopted a similar approach to evaluate the first order wave forces on closely spaced group of vertical cylinders. A series solution was obtained, the coefficients of which were determined forming a set of matrix equations. Results were given for two cylinders in various configurations relative to the incident wave angles. Chakrabarti (1978) later used the same method for the derivation of the velocity potential for any number of cylinders in the presence of linear waves. Simon (1982) proposed an approximate theory for computing wave forces on an array of wave energy devices. According to this theory, a diverging wave scattering from one cylinder is replaced by a plane wave of appropriate amplitude in the vicinity of another cylinder. Once the amplitude and phase of the equivalent plane wave have been determined, the problem reduces to summing the effects of the plane waves on any given cylinder. The effect of this equivalent plane wave on the given cylinder is then computed. The solution inherently assumes that the spacing between two cylinders is fairly large relative to the incident wavelength. More recently, McIver and Evans (1984) formulated a new matrix method based upon the idea due to Simon for the linear multicylinder problem. In this method, the efficiency is improved by approximating the effect on one cylinder of scattered waves from other cylinders as that of plane incident waves. Their numerical solution for three cylinders case gives good results even when the cylinders are close to one another.

The previous analysis for multiple scattering in arrays is restricted to the linearised (small amplitude) wave theory only. However, the hydrodynamic interactions caused by nonlinear (finite amplitude) wave theory are not well understood and to our knowledge, have not been reported in the literature. Most of the previous investigations in this area has been directed towards extending the linear theory of MacCamy and Fuchs (1954) for an isolated, bottom mounted, surface piercing cylindrical pile to include second order terms in the analysis (see Molin (1979), Eatock Taylor and Hung (1987) and Rahman (1984, 1987)). Recently, Masuda (1986, 1987) has presented a numerical technique to calculate the second order diffraction loads on arrays of vertical cylinders of arbitrary cross sections. The method needs a numerical integration over the entire mean fluid free surface and this is computationally very expensive.

In the present paper, a method is formulated to estimate the hydrodynamic forces to the second order on a pair of bottom mounted, surface piercing cylinders in water of arbitrary uniform depth. In this method, we have incorporated the idea initially presented by Simon where a diverging wave scattered from one cylinder is replaced by equivalent plane waves of appropriate amplitude and phase in the neighbourhood of another cylinder. Specifically we have employed this method in determining the resultant radiation potential at a given cylinder. In this paper the second order wave forces due to second order potential are calculated using an extension of Lighthill's method (1979) for finite water depth. Theoretical results are presented which illustrate the influence of the relative spacing between the cylinders and the incident wave frequency on the hydrodynamic loads to the second order. The computation of the second order wave forces on a single cylinder is discussed with some numerical results. Also the drift forces on an isolated cylinder were computed.

## 2 Mathematical formulation

In this paper we consider two right circular cylinders of radius $a$ separated by a distance $s$. We assume that these cylinders extend from the ocean bed to the free surface, as shown in Figure 1. A fixed coordinate system $0xyz$ is employed with the $x$ and $y$ axes in the horizontal plane with the $z$-axis pointing vertically upwards from an origin on the free surface. The centers of the two cylinders $0_1$ and $0_2$, lie on $0y$. The water depth is $h$. Long crested sinusoidal waves of frequency $\sigma$ and amplitude $A$ propagate in the positive $x$ direction. The linear velocity potential $\Phi$ may be written

$$\Phi = Re[\phi e^{i\sigma t}] \tag{1}$$

where $Re$ stands for real part of [ ]. The potential $\phi$ is conveniently decomposed into contributions $\phi_I$ and $\phi_S$ from incident and scattered waves respectively. In terms of cylindrical coordinates $r_1, \theta_1, z$, we may express $\phi_I$ relative to the center of cylinder 1 as

$$\phi_I(r_1,\theta_1,z) = \frac{gA}{\sigma}\frac{\cosh k(z+h)}{\cosh kh}\sum_{n=-\infty}^{\infty}(-i)^n J_n(kr_1)exp(in\theta_1). \qquad (2)$$

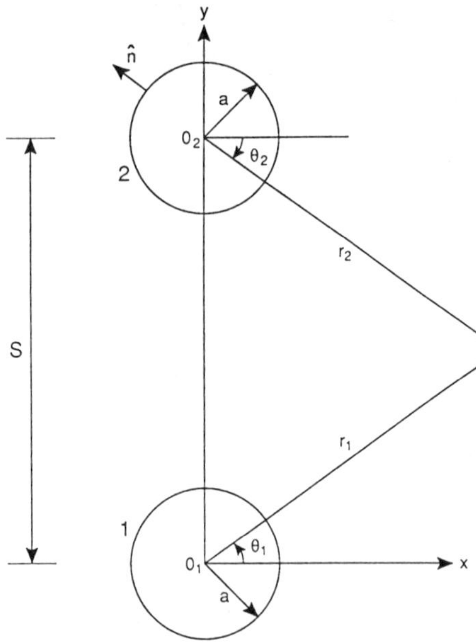


Figure 1: Coordinate system for two cylinders.

The far field scattered potential due to cylinder 1 may be written

$$\phi_{S1}(r_1,\theta_1,z) = (\frac{gA}{\sigma}\frac{\cosh k(z+h)}{\cosh kh})\sum_{n=-\infty}^{\infty} a_n(-i)^n H_n^{(2)}(kr_1)exp(in\theta_1) \qquad (3)$$

where $\sigma^2 = gk\tanh kh$ and $k$ is the wave number, $H_n^{(2)}$ is a Hankel function of the second kind satisfying the radiation condition, and the coefficients $a_n$ are to be determined from the boundary condition on the cylinder.

In view of the symmetry about $0x$, the scattered potential due to cylinder 2 must be expressible in terms of cylindrical coordinates $r_2, \theta_2, z$ in the same form :

$$\phi_{S2}(r_2,\theta_2,z) = (\frac{gA}{\sigma}\frac{\cosh k(z+h)}{\cosh kh})\sum_{n=-\infty}^{\infty} a_n(-i)^n H_n^{(2)}(kr_2)exp(in\theta_2). \qquad (4)$$

Equation (4) can be conveniently transferred into coordinates $r_1, \theta_1$ and $z$ using the Graf addition theorem for Bessel functions (Equation 9.1.79 given by Abramowitz and Stegun (1972)). Referring to the geometry of Figure 1, we can write

$$H_n^{(2)}(kr_2)exp[-in(\frac{\pi}{2}-\theta_2)] = \sum_{m=-\infty}^{\infty} H_{m+n}^{(2)}(ks)J_m(kr_1)exp[-im(\frac{\pi}{2}-\theta_1)]. \qquad (5)$$

Thus Equation (4) is transferred into

$$\phi_{S2}(r_1,\theta_1,z) = (\frac{gA}{\sigma}\frac{\cosh k(z+h)}{\cosh kh}) \sum_{n=-\infty}^{\infty} a_n \sum_{m=-\infty}^{\infty} (-i)^m H^{(2)}_{m+n}(ks) \times J_m(kr_1) exp(im\theta_1). \tag{6}$$

This then can be treated as the second scattered potential due to the cylinder 1 such that total potential is $\phi = \phi_I + \phi_{s1} + \phi_{s2}$. The body surface boundary condition at the cylinder 1 yields

$$\sum_{n=-\infty}^{\infty} (-i)^n J'_n(ka) exp(in\theta_1) + \sum_{n=-\infty}^{\infty} a_n (-i)^n H^{(2)\prime}_n(ka) exp(in\theta_1) + \sum_{n=-\infty}^{\infty} a_n \sum_{m=-\infty}^{\infty} (-i)^m H^{(2)}_{m+n}(ks) J'_m(ka) exp(im\theta_1) = 0 \tag{7}$$

which holds for all $\theta_1$ in the range $(0, 2\pi)$.
Rearranging this equation and interchanging the order of the two convergent summations, we obtain equations for $a_n$:

$$a_n \frac{H^{(2)\prime}_n(ka)}{J'_n(ka)} + \sum_{m=-\infty}^{\infty} a_m H^{(2)}_{m+n}(ks) = -1, \quad -\infty < n < \infty \tag{8}$$

where $n$ is an integer. Therefore the combined velocity potential $\phi = \phi_I + \phi_{s1} + \phi_{s2}$ expressed in the coordinates $r_1, \theta_1$ and $z$ is then given by

$$\phi(r_1,\theta_1,z) = (\frac{gA}{\sigma}\frac{\cosh k(z+h)}{\cosh kh}) \sum_{n=-\infty}^{\infty} (-i)^n exp(in\theta_1)[J_n(kr_1) + a_n H^{(2)}_n(kr_1) + \sum_{m=-\infty}^{\infty} a_m H^{(2)}_{m+n}(ks) J_n(kr_1)].$$

This corresponds to the incident wave plus the total effect of scattering by the two cylinders. But with the aid of (8), this may be written in the simple form:

$$\phi(r_1,\theta_1,z) = (\frac{gA}{\sigma}\frac{\cosh k(z+h)}{\cosh kh}) \sum_{n=-\infty}^{\infty} a_n [H^{(2)}_n(kr_1) - \frac{H^{(2)\prime}_n(ka)}{J'_n(ka)} J_n(kr_1)] \times exp(i(n\theta_1 - n\frac{\pi}{2})) \tag{9}$$

On the surface of cylinder 1, the expression for $\phi(a, \theta, z)$ becomes then

$$\phi(a,\theta,z) = (\frac{gA}{\sigma}\frac{\cosh k(z+h)}{\cosh kh}) \sum_{n=-\infty}^{\infty} \frac{a_n}{J'_n(ka)} exp[i(n\theta_1 - \frac{n}{2}\pi)] \frac{2i}{\pi ka}. \tag{10}$$

Equations (9) and (10) are the complex forms of the linear velocity potential at $(r_1, \theta_1, z)$ and $(a, \theta_1, z)$ respectively. In the following we shall find the first order forces. Lighthill's technique (1979) will be used to obtain the second order forces from the knowledge of linear potential $\Phi$ alone. The difficulty of determining these forces lies in the evaluation of the coefficients $a_n$.

# 3 First order wave forces

The first order forces on cylinder 1 can be determined from the linearized Bernoulli equation as

$$F_x^{(1)} = \int_{-h}^{0}\int_{0}^{2\pi}(-\rho\frac{\partial\Phi}{\partial t})(a\cos\theta)d\theta dz \quad (11)$$

$$F_y^{(1)} = \int_{-h}^{0}\int_{0}^{2\pi}(-\rho\frac{\partial\Phi}{\partial t})(a\sin\theta)d\theta dz \quad (12)$$

where $F_x^{(1)}$ and $F_y^{(1)}$ denote the components of horizontal force along $x$ and $y$ directions.

By substituting $\Phi$ into (11) and (12), we readily derive the exact first order linear forces as

$$F_x^{(1)} = -\frac{2\rho gA}{k^2 J_1'(ka)}\tanh kh Re[(a_1 + a_{-1})e^{i(\sigma t+\pi/2)}] \quad (13)$$

$$F_y^{(1)} = \frac{2\rho gA}{k^2 J_1'(ka)}\tanh kh Re[(a_1 - a_{-1})e^{i\sigma t}]. \quad (14)$$

For an isolated cylinder, Equation (8) reduces to

$$a_n\frac{H_n^{(2)\prime}(ka)}{J_n'(ka)} = -1$$

from which it is clear that $a_n = a_{-n}$ and hence $a_1 = a_{-1}$. Therefore, we obtain $F_y^{(1)}$ =0 and

$$F_x^{(1)} = C_M(\rho gA\pi a^2)\tanh kh\cos(\sigma t - \beta) \quad (15)$$

where

$$C_M = \frac{4}{(\pi k^2a^2)\sqrt{J_1'^2(ka) + Y_1'^2(ka)}} \quad (16)$$

$$\beta = \tan^{-1}(\frac{J_1'}{Y_1'}). \quad (17)$$

It is to be noted here that these results are the same as the previously published results (MacCamy and Fuchs (1954)).

# 4 Second order wave forces on a single cylinder

In this section we discuss a method to compute the quadratic forces on a vertical cylinder as initiated by Molin (1979) and later used by Eatock Taylor and Hung (1987).

Consider the total velocity potential

$$\Phi = \Phi_\ell + \Phi_q \quad (18)$$

where

$$\begin{aligned} \Phi_\ell &= Re(\phi_\ell e^{i\sigma t}) && (19)\\ \Phi_q &= Re(\phi_q e^{2i\sigma t}). && (20)\end{aligned}$$

Here $\Phi_\ell$ and $\Phi_q$ are, respectively, linear and quadratic potentials; $\phi_\ell$ and $\phi_q$ are complex linear and quadratic spatial potentials, respectively.
The second order force can be decomposed as :

$$F^{(2)} = F_m^{(2)} + F_0^{(2)} + F_q^{(2)} \tag{21}$$

where

$$\begin{aligned} F_m^{(2)} &= Re[\frac{\rho g}{4}\int_{z=0}\eta^{(1)}.\eta^{(1)^*}n_\beta dl - \frac{\rho}{4}\int_{S_0}\nabla\phi^{(1)}.\nabla\phi^{(1)^*}n_\beta dS] && (22)\\ F_0^{(2)} &= Re[\{\frac{\rho g}{4}\int_{z=0}\eta^{(1)}.\eta^{(1)}n_\beta dl - \frac{\rho}{4}\int_{S_0}\nabla\phi^{(1)}.\nabla\phi^{(1)}n_\beta dS\}e^{2i\sigma t}] && (23)\\ F_q^{(2)} &= Re[(-2i\rho\sigma)e^{2i\sigma t}\int\int_{S_0}\phi_q\frac{\partial\psi}{\partial n}ds]. && (24)\end{aligned}$$

Here $\psi$ is the radiation potential and $\eta^{(1)}$ is the first order wave elevation. The computation of $F_m^{(2)}$ and $F_0^{(2)}$ (mean and oscillatory part of dynamic and waterline forces respectively) is simple once the first order solution is found. The difficulty arises in the computation of the quadratic force $F_q^{(2)}$. Now define

$$\begin{aligned} \phi_\ell &= \phi_I^{(1)} + \phi_s^{(1)}\\ \phi_q &= \phi_I^{(2)} + \phi_s^{(2)} && (25)\end{aligned}$$

where $\phi_I^{(2)}$ and $\phi_s^{(2)}$ are second order incident and scattered propagating potentials respectively.
Therefore, Equation (24) can be written as

$$F_q^{(2)} = Re[(-2i\rho\sigma)\{\int\int_{S_0}\phi_I^{(2)}\frac{\partial\psi}{\partial n}ds + \int\int_{S_0}\phi_s^{(2)}\frac{\partial\psi}{\partial n}ds\}e^{2i\sigma t}]. \tag{26}$$

We now use Green's second identity on the surfaces enclosing $V$,

$$\int\int_{S_0US_BUS_FUS_\infty}(\phi_s^{(2)}\frac{\partial\psi}{\partial n} - \psi\frac{\partial\phi_s^{(2)}}{\partial n})ds = 0$$

which simply can be written as

$$\begin{aligned} \int\int_{S_0}\phi_s^{(2)}\frac{\partial\psi}{\partial n}ds &= \int\int_{S_0}\psi\frac{\partial\phi_s^{(2)}}{\partial n}ds + \int\int_{S_B}(\psi\frac{\partial\phi_s^{(2)}}{\partial n} - \phi_s^{(2)}\frac{\partial\psi}{\partial n})ds\\ &+ \int\int_{S_F}(\psi\frac{\partial\phi_s^{(2)}}{\partial n} - \phi_s^{(2)}\frac{\partial\psi}{\partial n})ds + \int\int_{S_\infty}(\psi\frac{\partial\phi_s^{(2)}}{\partial n} - \phi_s^{(2)}\frac{\partial\psi}{\partial n})ds.\end{aligned}$$

Now introducing body surface condition and bottom boundary condition in to this equation, the quadratic forces (Equation (26)) can be expressed as

$$F_q^{(2)} = Re[\{F_{qI}^{(2)} + F_{qII}^{(2)} + F_{qIII}^{(2)} + F_{qIV}^{(2)}\}e^{2i\sigma t}] \tag{27}$$

where

$$F_{qI}^{(2)} = (-2i\rho\sigma)\int\int_{S_0} \phi_I^{(2)} \frac{\partial \psi}{\partial n} ds \tag{28}$$

$$F_{qII}^{(2)} = 2i\rho\sigma \int\int_{S_0} \psi \frac{\partial \phi_I^{(2)}}{\partial n} ds \tag{29}$$

$$F_{qIII}^{(2)} = (-2i\rho\sigma)\int\int_{S_F} (\psi \frac{\partial \phi_s^{(2)}}{\partial n} - \phi_s^{(2)} \frac{\partial \psi}{\partial n}) ds \tag{30}$$

$$F_{qIV}^{(2)} = (-2i\rho\sigma)\int\int_{S_\infty} (\psi \frac{\partial \phi_s^{(2)}}{\partial n} - \phi_s^{(2)} \frac{\partial \psi}{\partial n}) ds. \tag{31}$$

Since the second order incident potential and the radiation potential can be obtained, computation of $F_{qI}^{(2)}$ and $F_{qII}^{(2)}$ is straightforward. Also $F_{qIV}^{(2)}$ decays to zero as $r \to \infty$. The evaluation of $F_{qIII}^{(2)}$ is discussed in the following. The free surface boundary condition must be satisfied by $\phi_I^{(2)}$ and $\phi_s^{(2)}$ in the following manner:

$$\begin{aligned} \frac{\partial \phi_I^{(2)}}{\partial z} - K\phi_I^{(2)} &= [-\frac{i\sigma}{g}(\nabla \phi_I^{(1)})^2 \\ &+ \frac{1}{2g}(i\sigma k^2) sech^2 kh (\phi_I^{(1)})^2]_{z=0} \end{aligned} \tag{32}$$

and

$$\begin{aligned} \frac{\partial \phi_s^{(2)}}{\partial z} - K\phi_s^{(2)} &= \{-\frac{i\sigma}{g}[(\nabla \phi_\ell)^2 - (\nabla \phi_I^{(1)})^2] \\ &+ \frac{i\sigma k^2}{2g} sech^2 kh [\phi_\ell^{(2)} - \phi_I^{(1)2}]\}_{z=0}. \end{aligned} \tag{33}$$

Now $F_{qIII}^{(2)}$ can be rewritten as

$$F_{qIII}^{(2)} = (-2i\rho\sigma)\int\int_{S_F} \psi(\frac{\partial \phi_s^{(2)}}{\partial z} - K\phi_s^{(2)}) ds$$

which can subsequently be reduced to the following

$$F_{qIII}^{(2)} = -\frac{2\sigma^2\rho}{g}\{\int\int_{S_F} \psi((\nabla\phi_\ell)^2 - (\nabla\phi_I^{(1)})^2 - \frac{k^2 sech^2 kh}{2}(\phi_\ell^2 - \phi_I^{(1)2}))ds\}. \tag{34}$$

Therefore the calculation of $F_{qIII}^{(2)}$ is possible with the knowledge of first order potentials alone.

The first order incident potential $\phi_I^{(1)}$ can be written as :

$$\phi_I^{(1)} = \sum_{n=0}^{\infty} \epsilon_n A_n \frac{\cosh k(z+h)}{\cosh kh} \cos n\theta \tag{35}$$

where $\epsilon_0 = 1$, $\epsilon_n = 2$ for $n \neq 0$,

$$A_n = \frac{gA}{\sigma}(-i)^n J_n(kr) = \hat{A}_n J_n(kr). \tag{36}$$

The corresponding scattered potential $\phi_S^{(1)}$ is given by :

$$\phi_S^{(1)} = \sum_{n=0}^{\infty} \epsilon_n B_n \frac{\cosh k(z+h)}{\cosh kh} \cos n\theta \tag{37}$$

where

$$B_n = -\frac{gA}{\sigma}(-i)^n \frac{J_n'(ka)}{H_n^{(2)'}(ka)} H_n^{(2)}(kr) = \hat{B}_n H_n^{(2)}(kr). \tag{38}$$

The time harmonic component of the second order incident potential $\phi_I^{(2)}$ is

$$\phi_I^{(2)} = \sum_{n=0}^{\infty} \epsilon_n D_n \cosh 2k(z+h) \cos n\theta \tag{39}$$

where

$$D_n = -\frac{3iA^2\sigma}{8\sinh^4 kh}(-i)^n J_n(2kr) = \hat{D}_n J_n(2kr). \tag{40}$$

The radiation potential corresponding to a frequency $2\sigma$ is :

$$\psi = \{E_0 \frac{\cosh \lambda(z+h)}{\cosh \lambda h} + \sum_{j=1}^{\infty} E_j \frac{\cos m_j(z+h)}{\cos m_j h}\} \cos\theta \tag{41}$$

where

$$\begin{aligned} E_0 &= \frac{2\sinh 2\lambda h}{2\lambda h + \sinh 2\lambda h} \frac{H_1^{(2)}(\lambda r)}{H_1^{(2)'}(\lambda a)} \\ &= \hat{E}_0 H_1^{(2)}(\lambda r) \\ E_j &= \frac{2\sin 2m_j h}{2m_j h + \sin 2m_j h} \frac{K_1(m_j r)}{K_1'(m_j a)} \end{aligned}$$

Here $K_1$ is the modified Bessel function of the second kind of order 1 and $\lambda$ and $m_j$ satisfy the corresponding equations respectively

$$\begin{aligned} 4\sigma^2 &= g\lambda \tanh \lambda h \\ 4\sigma^2 &= -gm_j \tan m_j h. \end{aligned}$$

After substitutions the force components become

$$F_m^{(2)} = Re[\pi\rho a\{\sum_{n=0}^{\infty}(-\frac{\sigma^2}{gk} + k^2 Z_1 + \frac{Z_2}{a^2}n(n+1))G_n G_{n+1}^*\}]_{r=a} \tag{42}$$

$$F_0^{(2)} = Re[\pi\rho a\{\sum_{n=0}^{\infty}(\frac{\sigma^2}{gk} + k^2 Z_1 + \frac{Z_2}{a^2}n(n+1))G_n G_{n+1}\}e^{2i\sigma t}]_{r=a} \tag{43}$$

$$F_{qI}^{(2)} = 4i\sigma\rho a\pi[D_1\sum_{j=0}^{\infty}E_j' Z_{3j}]_{r=a}, \tag{44}$$

$$F_{qII}^{(2)} = -4i\sigma\rho a\pi[D_1'\sum_{j=0}^{\infty}E_j Z_{3j}]_{r=a} \tag{45}$$

where

$$G_n = A_n + B_n,$$
$$Z_1 = \frac{\sinh 2kh - 2kh}{4k^2\cosh^2 kh},$$
$$Z_2 = \frac{\sinh 2kh + 2kh}{4k^2\cosh^2 kh},$$
$$Z_{30} = \frac{2k\sinh 2kh\cosh\lambda h - \lambda\cosh 2kh\sinh\lambda h}{4k^2-\lambda^2} sech\lambda h,$$
$$Z_{3j} = \frac{2k\sinh 2kh\cos m_j h + m_j\cosh 2kh\sin m_j h}{4k^2+m_j^2}\sec m_j h.$$

Now decomposing $F_{qIII}^{(2)}(a,\infty)$ into two parts, one $F_{qIII}^{(2)}(a,b)$ for inner region ($a < r < b$) and the other $F_{qIII}^{(2)}(b,\infty)$ for outer region ($b < r < \infty$) we get

$$F_{qIII}^{(2)}(a,b) = -\frac{8\rho\sigma^2\pi}{g}\int_a^b(\sum_{j=0}^{\infty}E_j)\sum_{n=0}^{\infty}\{(G_n' G_{n+1}' - A_n' A_{n+1}') + (\frac{\sigma^4}{g^2} - \frac{k^2 sech^2 kh}{2} + \frac{n(n+1)}{r^2})(G_n G_{n+1} - A_n A_{n+1})\}rdr \tag{46}$$

The use of asymptotic expansions of the potential functions, namely,

$$\phi_I^{(1)} \simeq \sum_{n=0}^{\infty}\epsilon_n\hat{A}_n\sqrt{\frac{2}{\pi kr}}\cos(kr-\alpha_n)\cos n\theta$$
$$\phi_S^{(1)} \simeq \sum_{n=0}^{\infty}\epsilon_n\hat{B}_n\sqrt{\frac{2}{\pi kr}}e^{-i(kr-\alpha_n)}\cos n\theta$$
$$\psi \simeq \hat{E}_0\sqrt{\frac{2}{\pi kr}}e^{-i(kr-\alpha_1)}\cos\theta \tag{47}$$

where $\alpha_n = \frac{2n+1}{4}\pi$, yields

$$F_{qIII}^{(2)}(b,\infty) = \frac{4\rho\sigma^2 k}{g}\int_b^{\infty}\hat{E}_0\sqrt{\frac{2}{\pi\lambda r}}e^{-i(\lambda r-\alpha_1)}$$

$$\times \sum_{n=0}^{\infty}[3sech^2kh(2\hat{B}_n\hat{B}_{n+1} + \hat{B}_n\hat{A}_{n+1} + \hat{A}_n\hat{B}_{n+1})e^{-i(2kr-\alpha_n-\alpha_{n+1})}$$
$$+(3sech^2kh - 4)\{\hat{B}_n\hat{A}_{n+1}e^{i(\alpha_n-\alpha_{n+1})} + \hat{A}_n\hat{B}_{n+1}e^{i(\alpha_{n+1}-\alpha_n)}\}]dr$$

Making use of Fresnel sine and cosine integrals this expression can be written as

$$\begin{aligned} F^{(2)}_{qIII}(b,\infty) &= \frac{8\rho\sigma^2 k}{g}\hat{E}_0\frac{e^{i\alpha_1}}{\sqrt{\lambda}}\sum_{n=0}^{\infty}[3sech^2kh \\ &\times(2\hat{B}_n\hat{B}_{n+1} + \hat{B}_n\hat{A}_{n+1} + \hat{A}_n\hat{B}_{n+1})e^{i(\alpha_n+\alpha_{n+1})} \\ &\times\sqrt{\frac{1}{K'}}[\{\frac{1}{2} - C(\sqrt{\frac{2K'b}{\pi}})\} - i\{\frac{1}{2} - S(\sqrt{\frac{2K'b}{\pi}})\}] \\ &+(3sech^2kh - 4)\{\hat{B}_n\hat{A}_{n+1}e^{i(\alpha_n-\alpha_{n+1})} + \hat{A}_n\hat{B}_{n+1}e^{i(\alpha_{n+1}-\alpha_n)}\} \\ &\times\sqrt{\frac{1}{\lambda}}[\{\frac{1}{2} - C(\sqrt{\frac{2\lambda b}{\pi}})\} - i\{\frac{1}{2} - S(\sqrt{\frac{2\lambda b}{\pi}})\}]] \end{aligned} \tag{48}$$

where $K' = 2k + \lambda$, and the Fresnel sine and cosine integrals are $S(t) = \int_0^t \sin u^2 du$ and $C(t) = \int_0^t \cos u^2 du$ respectively.

# 5 Second order wave forces on a pair of cylinders

Referring to the work of Lighthill (1979), we can determine the second order wave forces by considering the following three forces: (a) waterline force (b) dynamic force (c) quadratic force. We shall discuss these forces one after another. However, in this section we shall be concerned with the second order wave forces mainly due to the waterline force and the dynamic force. These two forces can easily be obtained from the knowledge of the first order linear potential given by Bernoulli's Equation. The horizontal components of the waterline forces can be determined from the formulae:

$$\begin{aligned} F^{(2)}_{wx} &= \int_0^{2\pi}(\frac{\rho}{2g})(\frac{\partial\Phi}{\partial t})^2_{\substack{r_1=a\\ z=0}}(-a\cos\theta)d\theta \\ &= (-\frac{\rho g A^2}{a\pi^2k^2})Re\int_0^{2\pi}[\phi^2 e^{2i\sigma t} + \phi\phi^*]\cos\theta d\theta \end{aligned} \tag{49}$$

$$\begin{aligned} F^{(2)}_{wy} &= \int_0^{2\pi}(\frac{\rho}{2g})(\frac{\partial\Phi}{\partial t})^2(-a\sin\theta)d\theta \\ &= (-\frac{\rho g A^2}{a\pi^2k^2})Re\int_0^{2\pi}[\phi^2 e^{2i\sigma t} + \phi\phi^*]\sin\theta d\theta. \end{aligned} \tag{50}$$

At $z = 0$ and $r_1 = a$, the expression for $\phi$ is

$$\phi = (\frac{2gA}{\pi k a\sigma})\sum_{n=-\infty}^{\infty}\frac{a_n}{J'_n(ka)}exp[-i(n\theta - (n-1)\frac{\pi}{2})] \tag{51}$$

By substituting Equation (51), and performing integration with respect to $\theta$ in (49) and (50), we may write these in the form

$$F_{wx}^{(2)} = (\frac{-\rho g A^2}{\pi a k^2}) Re[i e^{2i\sigma t} \sum_{n=-\infty}^{\infty} (\frac{a_n}{J_n'})(\frac{a_{-1-n}}{J_{-1-n}'} - \frac{a_{1-n}}{J_{1-n}'}) + i \sum_{n=-\infty}^{\infty} (\frac{a_n}{J_n'})(\frac{a_{n+1}^*}{J_{n+1}'} - \frac{a_{n-1}^*}{J_{n-1}'})] \quad (52)$$

$$F_{wy}^{(2)} = (\frac{-\rho g A^2}{\pi a k^2}) Re[e^{2i\sigma t} \sum_{n=-\infty}^{\infty} (\frac{a_n}{J_n'})(\frac{a_{-1-n}}{J_{-1-n}'} + \frac{a_{1-n}}{J_{1-n}'}) + \sum_{n=-\infty}^{\infty} (\frac{a_n}{J_n'})(\frac{a_{n+1}^*}{J_{n+1}'} + \frac{a_{n-1}^*}{J_{n-1}'})] \quad (53)$$

Similarly, the horizontal components of the dynamic forces may be obtained from the following formulae:

$$F_{dx}^{(2)} = \int_0^{2\pi} \{\int_{-h}^{0} \frac{-\rho}{2} (\nabla \Phi)^2 dz\}(-a \cos \theta) d\theta \quad (54)$$

$$F_{dy}^{(2)} = \int_0^{2\pi} \{\int_{-h}^{0} \frac{-\rho}{2} (\nabla \Phi)^2 dz\}(-a \sin \theta) d\theta \quad (55)$$

Performing the indicated integrations in (54) and (55), we may write these as follows:

$$\begin{aligned} F_{dx}^{(2)} = & (\frac{\rho g A^2}{2\pi a k^2})[\, Re\{ \sum_{n=-\infty}^{\infty} (1 - \frac{2kh}{\sinh 2kh})(\frac{a_n}{J_n'})(\frac{a_{1-n}}{J_{1-n}'} - \frac{a_{-1-n}}{J_{-1-n}'}) \\ & + (1 + \frac{2kh}{\sinh 2kh})(\frac{a_n}{J_n'})(\frac{n(n-1)}{k^2 a^2}\frac{a_{1-n}}{J_{1-n}'} - \frac{n(n+1)}{k^2 a^2}\frac{a_{-1-n}}{J_{-1-n}'}) i e^{2i\sigma t}\} \\ & + Re\{i \sum_{n=-\infty}^{\infty} (1 - \frac{2kh}{\sinh 2kh})(\frac{a_n}{J_n'})(\frac{a_{n-1}^*}{J_{n-1}'} - \frac{a_{n+1}^*}{J_{n+1}'}) \\ & + (1 + \frac{2kh}{\sinh 2kh})(\frac{a_n}{J_n'})(\frac{n(n+1)}{K^2 a^2}\frac{a_{n+1}^*}{J_{n+1}'} - \frac{n(n-1)}{k^2 a^2}\frac{a_{n-1}^*}{J_{n-1}'})\}] \end{aligned} \quad (56)$$

$$\begin{aligned} F_{dy}^{(2)} = & (\frac{-\rho g A^2}{2\pi a k^2})[\, Re\{ \sum_{n=-\infty}^{\infty} (1 - \frac{2kh}{\sinh 2kh})(\frac{a_n}{J_n'})(\frac{a_{1-n}}{J_{1-n}'} + \frac{a_{-1-n}}{J_{-1-n}'}) \\ & + (1 + \frac{2kh}{\sinh 2kh})(\frac{a_n}{J_n'})(\frac{n(n+1)}{k^2 a^2}\frac{a_{-1-n}}{J_{-1-n}'} - \frac{n(n-1)}{k^2 a^2}\frac{a_{1-n}}{J_{1-n}'}) e^{2i\sigma t}\} \\ & + Re[\sum_{n=-\infty}^{\infty} (1 - \frac{2kh}{\sinh 2kh})(\frac{a_n}{J_n'})(\frac{a_{n-1}^*}{J_{n-1}'} + \frac{a_{n+1}^*}{J_{n+1}'}) \\ & - (1 + \frac{2kh}{\sinh 2kh})(\frac{a_n}{J_n'})(\frac{n(n-1)}{k^2 a^2}\frac{a_{n-1}^*}{J_{n-1}'} + \frac{n(n+1)}{k^2 a^2}\frac{a_{n+1}^*}{J_{n+1}'})]] \end{aligned} \quad (57)$$

It may be noted here that Equations (52), (53), (56) and (57) contain the oscillatory and steady-state parts of the solution. Eatock Taylor and Hung (1985) obtained the steady-state part of the solution in the form of wave drift enhancement effects in the multicolumn structures.

# 6 Quadratic force calculations

Now we evaluate the loading due to the second order potential using Lighthill's method. This method does not require the explicit calculation of the second order potentials in order to obtain the second order forces due to that potential. Instead, the force components are obtained by solving a linear radiation potential problem at the double frequency $2\sigma$. This radiation potential in turn is linked to the required force expressions through the application of Green's identity. We can calculate the quadratic forces using the following formulae:

The horizontal quadratic force components can be obtained as follows:

$$\begin{aligned} F_{qx}^{(2)} &= \iint_{S_0} (-\rho \frac{\partial \Phi_q}{\partial t}) n_x ds \\ &= Re[(-2\rho i \sigma) e^{2i\sigma t} \iint_{S_0} \phi_q n_x ds] \end{aligned} \tag{58}$$

and

$$\begin{aligned} F_{qy}^{(2)} &= \iint_{S_0} (-\rho \frac{\partial \Phi_q}{\partial t}) n_y ds \\ &= Re[(-2\rho i \sigma) e^{2i\sigma t} \iint_{S_0} \phi_q n_y ds] \end{aligned} \tag{59}$$

or in compact form

$$F_{qn}^{(2)} = Re[(-2\rho i \sigma) e^{2i\sigma t} \iint_{S_0} \phi_q \vec{n} ds] \tag{60}$$

where $\vec{n} = (n_x, n_y)$ are unit normal vector components as shown in Figure 1.

Next we introduce a radiation potential due to unit translation of the body with $2\sigma$ frequency as follows:

$$\Psi = Re(\psi e^{2i\sigma t}) \tag{61}$$

where $\psi$ satisfies the following linear boundary value problem (BVP)

$$\nabla^2 \psi = 0 \quad \text{in the equilibrium fluid region } V \tag{62}$$

$$\frac{\partial \psi}{\partial z} - \frac{4\sigma^2}{g} \psi = 0 \quad \text{in the equilibrium free surface } S_F \tag{63}$$

$$\frac{\partial \psi}{\partial n} = \begin{matrix} n_x \\ n_y \end{matrix} \quad \text{on the body } S_0 \tag{64}$$

$$\frac{\partial \psi}{\partial z} = 0 \quad \text{on the bottom } S_B \tag{65}$$

$\psi$ also satisfies the following Sommerfeld radiation condition

$$\lim_{r \to \infty} \sqrt{r} (\frac{\partial \psi}{\partial r} + i \lambda \psi) = 0 \qquad \text{at } S_\infty \tag{66}$$

where $\lambda$ is the second order wave number given by $4\sigma^2 = g\lambda \tanh \lambda h$.

Applying Green's second identity on equilibrium boundaries, we have

$$\iiint_V (\psi_q \nabla^2 \psi - \psi \nabla^2 \phi_q) dV = \iint_{S_0 U S_F U S_B U S_\infty} (\phi_q \frac{\partial \psi}{\partial n} - \psi \frac{\partial \phi_q}{\partial n}) ds.$$

Since $\nabla^2\psi = 0$ and $\nabla^2\phi_q = 0$, we have

$$\iiint_{S_0 U S_F U S_B U S_\infty} (\phi_q \frac{\partial \psi}{\partial n} - \psi \frac{\partial \phi_q}{\partial n}) ds = 0$$

which can be written as

$$\begin{aligned}\iint_{S_0} (\phi_q \frac{\partial \psi}{\partial n} \quad - \quad \psi \frac{\partial \phi_q}{\partial n}) ds &= \iint_{S_F} (\psi \frac{\partial \phi_q}{\partial n} - \phi_q \frac{\partial \psi}{\partial n}) ds \\ &+ \iint_{S_B} (\psi \frac{\partial \phi_q}{\partial n} - \phi_q \frac{\partial \psi}{\partial n}) ds + \iint_{S_\infty} (\psi \frac{\partial \phi_q}{\partial n} - \phi_q \frac{\partial \psi}{\partial n}) ds\end{aligned}$$

Since on the body $\frac{\partial \phi_q}{\partial n} = 0$ and $\frac{\partial \psi}{\partial n} = \begin{Bmatrix} n_x \\ n_y \end{Bmatrix}$ and on the ocean bottom $\frac{\partial \phi_q}{\partial n} = 0$ and $\frac{\partial \psi}{\partial n} = 0$. Substituting these values we obtain

$$\begin{aligned}\iint_{S_0} \phi_q \begin{Bmatrix} n_x \\ n_y \end{Bmatrix} ds &= \iint_{S_F} (\psi \frac{\partial \phi_q}{\partial n} - \phi_q \frac{\partial \psi}{\partial n}) ds \\ &+ \iint_{S_\infty} (\psi \frac{\partial \phi_q}{\partial n} - \phi_q \frac{\partial \psi}{\partial n}) ds.\end{aligned}$$

We can show that as $r \to \infty$

$$\iint_{S_\infty} (\psi \frac{\partial \phi_q}{\partial n} - \phi_q \frac{\partial \psi}{\partial n}) ds$$

oscillates and decays to zero as $r^{-1/2}$.
Thus we have

$$\iint_{S_0} \phi_q \begin{Bmatrix} n_x \\ n_y \end{Bmatrix} ds = \iint_{S_F} (\psi \frac{\partial \phi_q}{\partial n} - \phi_q \frac{\partial \psi}{\partial n}) ds. \tag{67}$$

Use of the free surface condition for $\psi$ yields

$$\iint_{S_0} \phi_q n ds = \iint_{S_F} \psi (\frac{\partial \phi_q}{\partial n} - K \phi_q) ds \tag{68}$$

where

$$K = \frac{4\sigma^2}{g}.$$

Now $F_{qn}^{(2)}$ becomes

$$F_{qn}^{(2)} \quad = \quad Re[(-2i\rho\sigma) \frac{e^{2i\sigma t}}{K} \iint_{S_F} W (\frac{\partial \phi_q}{\partial n} - K \phi_q) ds] \tag{69}$$

where $W = \frac{\partial \psi}{\partial z}|_{z=0}$ is the vertical velocity of the free surface.
To simplify $\iint_{S_F} W(\frac{\partial \phi_q}{\partial n} - K\phi_q) ds$, we present the following analysis:
We know that the free surface boundary condition can be written as

$$\begin{aligned}\frac{\partial^2 \Phi_q}{\partial t^2} + g \frac{\partial \Phi_q}{\partial z} &= -\frac{\partial}{\partial t}((\nabla \Phi_\ell)^2) \\ &+ \frac{1}{g}(\frac{\partial \Phi_\ell}{\partial t})[\frac{\partial}{\partial z}(\frac{\partial^2 \Phi_\ell}{\partial t^2} + g \frac{\partial \Phi_\ell}{\partial z})] \quad \text{at} \quad z = 0.\end{aligned} \tag{70}$$

Using the relations $\Phi_\ell = Re(\phi_\ell e^{i\sigma t})$ and $\Phi_q = Re(\phi_q e^{2i\sigma t})$ free surface condition becomes

$$\begin{aligned} g \ Re[(\frac{\partial \phi_q}{\partial z} - K\phi_q)e^{2i\sigma t}] &= -Re[(i\sigma)(\nabla\phi_\ell)^2 e^{2i\sigma t}] \\ &+ \frac{1}{2g} Re[(i\sigma\phi_\ell)(-\sigma^2\frac{\partial \phi_\ell}{\partial z} + g\frac{\partial^2 \phi_\ell}{\partial z^2})e^{2i\sigma t} \\ &+ (i\sigma\phi_\ell)^*(-\sigma^2\frac{\partial \phi_\ell}{\partial z} + g\frac{\partial^2 \phi_\ell}{\partial z^2})]. \end{aligned} \tag{71}$$

The last term on the right-hand side in this equation is purely imaginary. Hence it follows that

$$\begin{aligned} \frac{\partial \phi_q}{\partial z} - K\phi_q &= -\frac{1}{g}[(i\sigma)(\nabla\phi_\ell)^2] \\ &+ \frac{1}{2g^2}[(i\sigma\phi_\ell)(-\sigma^2\frac{\partial \phi_\ell}{\partial z} + g\frac{\partial^2 \phi_\ell}{\partial z^2})] \end{aligned} \tag{72}$$

which is to be evaluated at $z = 0$.

Thus the problem to find quadratic forces reduces to

$$\begin{aligned} F_{qn}^{(2)} &= Re[(-2i\rho\sigma)\frac{e^{2i\sigma t}}{K}\int\int_{S_F} W\{-\frac{1}{g}(i\sigma)(\nabla\phi_\ell)^2 \\ &+ \frac{1}{2g^2}(i\sigma\phi_\ell)(-\sigma^2\frac{\partial \phi_\ell}{\partial z} + g\frac{\partial^2 \phi_\ell}{\partial z^2})_{z=0}\}ds]. \end{aligned} \tag{73}$$

In our problem, however, the linear velocity potential on the cylinder 1 can be expressed as

$$\Phi_\ell(a,\theta,z,t) = \frac{2gA}{\pi ka\sigma}\frac{\cosh k(z+h)}{\cosh kh} \times Re[\sum_{n=-\infty}^{\infty}\frac{a_n}{J_n'}e^{i(n\theta+\sigma t-(n-1)\pi/2)}] \tag{74}$$

and so the complex potential is written as

$$\phi(a,\theta,z) = \frac{2gA}{\pi ka\sigma}(\frac{\cosh k(z+h)}{\cosh kh})\sum_{n=-\infty}^{\infty}\frac{a_n}{J_n'}e^{i(n\theta-(n-1)\pi/2)}. \tag{75}$$

Equations (74) and (75) represent the total velocity potentials (only propagating modes but not evanescent modes) on the cylinder 1 corresponding to the incident waves plus scattering effects by the cylinder 2 upon the cylinder 1. These expressions are evaluated at the surface of the cylinder 1.

The expression for the velocity potential for any $r$ can be written as

$$\Phi_\ell(r,\theta,z,t) = (\frac{gA}{\sigma}\frac{\cosh k(z+h)}{\cosh kh}) \times Re[\sum_{n=-\infty}^{\infty} a_n A_n(kr)e^{i(n\theta+\sigma t-n\pi/2)}] \tag{76}$$

and

$$\phi_\ell(r,\theta,z) = (\frac{gA}{\sigma}\frac{\cosh k(z+h)}{\cosh kh})\sum_{n=-\infty}^{\infty} a_n A_n(kr)e^{i(n\theta-n\pi/2)} \tag{77}$$

where

$$A_n(kr) = H_n^{(2)}(kr) - \frac{H^{(2)\prime}(ka)}{J_n'(ka)} J_n(kr). \tag{78}$$

At $z = 0$, it can be written as

$$\phi_\ell(r, 0, \theta) = (\frac{gA}{\sigma}) \sum_{n=-\infty}^{\infty} a_n A_n(kr) e^{i(n\theta - n\pi/2)}.$$

Thus taking the first and second derivatives of $\phi_\ell$ with respect to $z$ and putting $z = 0$ we have

$$(i\sigma\phi_\ell)(-\sigma^2 \frac{\partial \phi_\ell}{\partial z} + g\frac{\partial^2 \phi_\ell}{\partial z^2}) \quad = \quad (i\sigma k^2 g)(1 - \tanh^2 kh)\phi_\ell^2.$$

Therefore, Equation (73) can be simplified to yield

$$F_{qn}^{(2)} = Re[-\frac{\rho}{2} e^{2i\sigma t} \int\int_{S_F} W[(\nabla\phi_\ell)^2 - \frac{k^2}{2} sech^2 kh \ \phi_\ell^2] ds]. \tag{79}$$

# 7 Solution of linear radiation problem

In this section we consider the linear radiation boundary value problem in which the radiation potential $\Psi = Re(\psi e^{2i\sigma t})$ satisfies the Equation (62) subject to conditions (63) to (66). The solution can be written in the form

$$\psi = [E(z)H_1^{(2)}(\lambda r) + \sum_{j=1}^{\infty} F_j(z) K_1(m_j r)] \begin{pmatrix} \cos\theta \\ \sin\theta \end{pmatrix} \tag{80}$$

where

$$E(z) \quad = \quad \frac{2\cosh\lambda(z+h)\sinh\lambda h}{\lambda H_1^{(2)\prime}(\lambda a)(\lambda h + \sinh\lambda h\cosh\lambda h)} \tag{81}$$

$$F_j(z) \quad = \quad \frac{2\cos m_j(z+h)\sin m_j h}{m_j K_1'(m_j a)(m_j h + \sin m_j h \cos m_j h)} \tag{82}$$

Here $H_1^{(2)}$ and $K_1$ are the Hankel function and the modified Bessel function respectively.

The radiated potentials with respect to cylinder 1 and cylinder 2 then may be written respectively as

$$\psi_1(r_1, \theta_1, z) \quad = \quad [E(z)H_1^{(2)}(\lambda r_1) + \sum_{j=1}^{\infty} F_j(z) K_1(m_j r_1)] \begin{pmatrix} \cos\theta_1 \\ \sin\theta_1 \end{pmatrix} \tag{83}$$

$$\psi_2(r_2, \theta_2, z) \quad = \quad [E(z)H_1^{(2)}(\lambda r_2) + \sum_{j=1}^{\infty} F_j(z) K_1(m_j r_2)] \begin{pmatrix} \cos\theta_2 \\ \sin\theta_2 \end{pmatrix} \tag{84}$$

Thus in the vicinity of cylinder 1, the radiated potential $\psi_2$ may be written as (after using Graf's addition theorem)

$$\begin{aligned}\psi_2 &= [E(z)\sum_{m=-\infty}^{\infty}(-i)^{m-1}H_{m+1}^{(2)}(\lambda s)J_m(\lambda r_1)e^{im\theta_1}\\ &\quad+\sum_{j=1}^{\infty}F_j(z)\sum_{m=-\infty}^{\infty}K_{m+1}(m_js)I_m(m_jr_1)e^{im\theta_1}]\end{aligned} \tag{85}$$

Because we know from Graf's theorem that

$$\begin{aligned}H_1^{(2)}(\lambda r_2)e^{i\theta_2} &= \sum_{m=-\infty}^{\infty}(-i)^{m-1}H_{m+1}^{(2)}(\lambda s)J_m(\lambda r_1)e^{im\theta_1}\\ K_1(m_jr_2)e^{i\theta_2} &= \sum_{m=-\infty}^{\infty}(-i)^{m-1}K_{m+1}(m_js)I_m(m_jr_1)e^{im\theta_1}\end{aligned}$$

The combined radiated potential expressed in the coordinates $(r_1,\theta_1)$ is then given by

$$\begin{aligned}\psi &= \psi_1+\psi_2\\ &= [E(z)H_1^{(2)}(\lambda r_1)+\sum_{j=1}^{\infty}F_j(z)K_1(m_jr_1)]e^{i\theta_1}\\ &+ [E(z)\sum_{m=-\infty}^{\infty}(-i)^{m-1}H_{m+1}^{(2)}(\lambda r_1)e^{im\theta_1}\\ &\quad+\sum_{j=1}^{\infty}F_j(z)\sum_{m=-\infty}^{\infty}(-i)^{m-1}K_{m+1}(m_js)I_m(m_jr_1)e^{im\theta_1}]\end{aligned} \tag{86}$$

Let us assume $(r_1,\theta_1)=(r,\theta)$, then the $x$ and $y$ component of $\psi$ may be written respectively as

$$\begin{aligned}\psi_x &= [E(z)H_1^{(2)}(\lambda r)+\sum_{j=1}^{\infty}F_j(z)K_1(m_jr)]\cos\theta\\ &+ [E(z)\sum_{m=-\infty}^{\infty}(-i)^{m-1}H_{m+1}^{(2)}(\lambda s)J_m(\lambda r)\cos m\theta\\ &\quad+\sum_{j=1}^{\infty}F_j(z)\sum_{m=-\infty}^{\infty}(-i)^{m-1}K_{m+1}(m_js)I_m(m_jr)\cos m\theta]\end{aligned} \tag{87}$$

$$\begin{aligned}\psi_y &= [E(z)H_1^{(2)}(\lambda r)+\sum_{j=1}^{\infty}F_j(z)K_1(m_jr)\sin\theta]\\ &+ [E(z)\sum_{m=-\infty}^{\infty}(-i)^{m-1}H_{m+1}^{(2)}(\lambda s)J_m(\lambda r)\sin m\theta\\ &\quad+\sum_{j=1}^{\infty}F_j(z)\sum_{m=-\infty}^{\infty}(-i)^{m-1}K_{m+1}(m_js)I_m(m_jr)\sin m\theta]\end{aligned} \tag{88}$$

Here $\psi_x$ and $\psi_y$ may be defined as the potential due to surge and sway motion respectively. With the knowledge of $\psi$ we can then proceed to find the quadratic force components along $x$ and $y$ directions as follows:

$$\begin{aligned} F_{qx}^{(2)} &= Re[-\frac{\rho}{2}e^{2i\sigma t}\int_a^\infty rdr\int_0^{2\pi}(\frac{\partial\psi_x}{\partial z}) \\ &\quad [(\nabla\phi_\ell)^2+\frac{k^2}{2}(\tanh^2 kh-1)\phi_\ell^2]d\theta] \end{aligned} \tag{89}$$

$$\begin{aligned} F_{qy}^{(2)} &= Re[-\frac{\rho}{2}e^{2i\sigma t}\int_a^\infty rdr\int_0^{2\pi}(\frac{\partial\psi_y}{\partial z}) \\ &\quad [(\nabla\phi_\ell)^2+\frac{k^2}{2}(\tanh^2 kh-1)\phi_\ell^2]d\theta] \end{aligned} \tag{90}$$

The value of the integral

$$\begin{aligned} &\int_0^{2\pi}\ [(\nabla\phi_\ell)^2+\frac{k^2}{2}(\tanh^2 kh-1)\phi_\ell^2]\cos\theta d\theta \\ &= (\frac{g^2A^2k^2\pi}{\sigma^2})[\sum_{n=-\infty}^{\infty} ia_na_{-1-n}\{A_n'A_{-1-n}' \\ &+(\frac{n(n+1)}{k^2r^2}+\frac{3}{2}\tanh^2 kh-\frac{1}{2})A_nA_{-1-n}\} \\ &-ia_na_{-1-n}\{A_n'A_{1-n}'+(\frac{n(n-1)}{k^2r^2}+\frac{3}{2}\tanh^2 kh-\frac{1}{2})A_nA_{1-n}\}] \end{aligned} \tag{91}$$

Similarly the value of the integral

$$\begin{aligned} &\int_0^{2\pi}\ [(\nabla\phi_\ell)^2+\frac{k^2}{2}(\tanh^2 kh-1)\phi_\ell^2]\sin\theta d\theta \\ &= (\frac{g^2A^2k^2\pi}{\sigma^2})[\sum_{n=-\infty}^{\infty} a_na_{-1-n}\{A_n'A_{-1-n}' \\ &+(\frac{n(n+1)}{k^2r^2}+\frac{3}{2}\tanh^2 kh-\frac{1}{2})A_nA_{-1-n}\} \\ &+a_na_{1-n}\{A_n'A_{1-n}'+(\frac{n(n-1)}{k^2r^2}+\frac{3}{2}\tanh^2 kh-\frac{1}{2})A_nA_{1-n}\}] \end{aligned} \tag{92}$$

The integration with respect to $\theta$ in Equation (87) can be written as

$$\begin{aligned} &\int_0^{2\pi}\ (\frac{\partial\psi_x}{\partial z})_{z=0}\{(\nabla\phi_\ell)^2+\frac{k^2}{2}(\tanh^2 kh-1)\phi_\ell^2\}d\theta \\ &= \int_0^{2\pi}[E_0'H_1^{(2)}(\lambda r)+\sum_{j=1}^{\infty}F_{j0}'K_1(m_jr)] \\ &\quad [(\nabla\phi_\ell)^2+\frac{k^2}{2}(\tanh^2 kh-1)\phi_\ell^2]\cos\theta d\theta \\ &+ \int_0^{2\pi}[E_0'\sum_{\ell=-\infty}^{\infty}(-i)^{\ell-1}H_{\ell+1}^{(2)}(\lambda s)J_\ell(\lambda r) \end{aligned}$$

$$+\sum_{j=1}^{\infty} F'_{j0} \sum_{\ell=-\infty}^{\infty} (-i)^{\ell-1} K_{\ell+1}(m_j s) I_\ell(m_j r)]$$

$$[(\nabla\phi_\ell)^2 + \frac{k^2}{2}(\tanh^2 kh - 1)\phi_\ell^2] \cos \ell\theta d\theta \tag{93}$$

We know that

$$\begin{aligned}(\nabla\phi_\ell)^2 &+ \frac{k^2}{2}(\tanh^2 kh - 1)\phi_\ell^2 \\ &= (\frac{g^2 A^2 k^2}{\sigma^2}) \sum_{n=-\infty}^{\infty} \sum_{m=-\infty}^{\infty} a_n a_m \{A'_n A'_m \\ &+ (-\frac{mn}{k^2 r^2} + \frac{3}{2}\tanh^2 kh - \frac{1}{2}) A_n A_m\} \\ &e^{i((n+m)\theta - (n+m)\frac{\pi}{2})}\end{aligned} \tag{94}$$

Considering the second integral of Equation (91), we have after integration with respect to $\theta$,

$$\begin{aligned}\text{2nd integral } &= (\frac{\mathrm{g^2 k^2 A^2}}{\sigma^2}) \sum_{\mathrm{n}=-\infty}^{\infty} \sum_{\mathrm{m}=-\infty}^{\infty} \mathrm{a_n a_m}\{\mathrm{A'_n A'_m} + (-\frac{\mathrm{mn}}{\mathrm{k^2 r^2}} + \frac{3}{2}\tanh^2 \mathrm{kh} \\ &-\frac{1}{2}) A_n A_m\} \times \{(i\pi) E'_0 (H^{(2)}_{-n-m+1}(\lambda s) J_{-n-m}(\lambda r) + H^{(2)}_{n+m+1}(\lambda s) J_{n+m}(\lambda r)) \\ &+ \sum_{j=1}^{\infty} (i\pi) F'_{j0} (K_{-n-m+1}(m_j s) I_{-n-m}(m_j r) + K_{n+m+1} \\ &I_{n+m}(m_j r))\}\end{aligned} \tag{95}$$

Similarly the integration with respect to $\theta$ in Equation (88) can be written as

$$\begin{aligned}\int_0^{2\pi} & (\frac{\partial\psi_y}{\partial z})_{z=0} \{(\nabla\phi_\ell)^2 + \frac{k^2}{2}(\tanh^2 kh - 1)\phi_\ell^2\} d\theta \\ &= \int_0^{2\pi} [E'_0 H_1^{(2)}(\lambda r) + \sum_{j=1}^{\infty} F'_{j0} K_1(m_j r)] \\ &[(\nabla\phi_\ell)^2 + \frac{k^2}{2}(\tanh^2 kh - 1)\phi_\ell^2] \sin\theta d\theta \\ + & \int_0^{2\pi} [E'_0 \sum_{\ell=-\infty}^{\infty} (-i)^{\ell-1} H^{(2)}_{\ell+1}(\lambda s) J_\ell(\lambda r) \\ &+ \sum_{j=1}^{\infty} F'_{j0} \sum_{\ell=-\infty}^{\infty} (-i)^{\ell-1} K_{\ell+1}(m_j s) I_\ell(m_j r)] \\ &\times [(\nabla\phi_\ell)^2 + \frac{k^2}{2}(\tanh^2 kh - 1)\phi_\ell^2] \sin \ell\theta d\theta\end{aligned} \tag{96}$$

Considering the second integral of the Equation (92), we obtain after integration with respect to $\theta$:

$$\text{2nd integral } = (\frac{\mathrm{g^2 k^2 A^2}}{\sigma^2}\pi) \sum_{\mathrm{n}=-\infty}^{\infty} \sum_{\mathrm{m}=-\infty}^{\infty} \mathrm{a_n a_m}\{\mathrm{A'_n A'_m} + (-\frac{\mathrm{mn}}{\mathrm{k^2 r^2}} + \frac{3}{2}\tanh^2 \mathrm{kh}$$

$$-\frac{1}{2})A_nA_m\} \times \{E_0'(H^{(2)}_{-n-m+1}(\lambda s)J_{-n-m}(\lambda r) - (-1)^{n+m}H^{(2)}_{n+m+1}(\lambda s)$$
$$J_{n+m}(\lambda r)) + \sum_{j=1}^{\infty} F_{j0}'(K_{-n-m+1}(m_js)I_{-n-m}(m_jr)$$
$$-(-1)^{m+n}K_{n+m+1}I_{n+m}(m_jr))\} \tag{97}$$

Here $I_n$ and $K_n$ are the modified Bessel functions of first and second kind, and a prime denotes the differential with respect to the argument.

Thus combining this information, we can write the quadratic forces in the following manner:

$$F^{(2)}_{qx} = Re[-\frac{\rho}{2}e^{2i\sigma t}\int_a^{\infty} I_x r dr] \tag{98}$$
$$F^{(2)}_{qy} = Re[-\frac{\rho}{2}e^{2i\sigma t}\int_a^{\infty} I_y r dr] \tag{99}$$

where

$$I_x = (\frac{g^2A^2k^2\pi}{\sigma^2}[(E_0'H_1^{(2)}(\lambda r) + \sum_{j=1}^{\infty} F_{j0}'K_1(m_jr))$$
$$\times \sum_{n=-\infty}^{\infty}[ia_na_{-1-n}\{A_n'A_{-1-n}' + (\frac{n(n-1)}{k^2r^2} + \frac{3}{2}\tanh^2 kh - \frac{1}{2})A_nA_{-1-n}\}$$
$$- ia_na_{1-n}\{A_n'A_{1-n}' + (\frac{n(n-1)}{k^2r^2} + \frac{3}{2}\tanh^2 kh - \frac{1}{2})A_nA_{1-n}\}]]$$
$$+ (\frac{g^2A^2k^2\pi}{\sigma^2})\sum_{n=-\infty}^{\infty}\sum_{m=-\infty}^{\infty} a_na_m\{A_n'A_m' + (\frac{-nm}{k^2r^2} + \frac{3}{2}\tanh^2 kh$$
$$-\frac{1}{2})A_nA_{mn}\} \times \{iE_0'(H^{(2)}_{-n-m+1}(\lambda s)J_{-n-m}(\lambda r) + H^{(2)}_{n+m+1}(\lambda s)J_{n+m}(\lambda r))$$
$$+\sum_{j=1}^{\infty} iF_{j0}'(K_{-n-m+1}(m_js)I_{-n-m}(m_jr) + K_{n+m+1}(m_js)$$
$$I_{n+m}(m_jr))\} \tag{100}$$

Similarly,

$$I_y = (\frac{g^2A^2k^2\pi}{\sigma^2}[(E_0'H_1^{(2)}(\lambda r) + \sum_{j=1}^{\infty} F_{j0}'K_1(m_jr))$$
$$\times \sum_{n=-\infty}^{\infty} a_na_{-1-n}\{A_n'A_{-1-n}' + (\frac{n(n-1)}{k^2r^2} + \frac{3}{2}\tanh^2 kh - \frac{1}{2})A_nA_{-1-n}\}$$
$$+ a_na_{1-n}\{A_n'A_{1-n}' + (\frac{n(n-1)}{k^2r^2} + \frac{3}{2}\tanh^2 kh - \frac{1}{2})A_nA_{1-n}\}]$$
$$+ (\frac{g^2A^2k^2\pi}{\sigma^2})\sum_{n=-\infty}^{\infty}\sum_{m=-\infty}^{\infty} a_na_m\{A_n'A_m' + (\frac{-nm}{k^2r^2} + \frac{3}{2}\tanh^2 kh$$
$$-\frac{1}{2})A_nA_{mn}\} \times \{E_0'(H^{(2)}_{-n-m+1}(\lambda s)J_{-n-m}(\lambda r) - (-1)^{n+m}H^{(2)}_{n+m+1}(\lambda s)$$

$$J_{n+m}(\lambda r)) + \sum_{j=1}^{\infty} F'_{j0}(K_{-n-m+1}(m_j s) I_{-n-m}(m_j r) - (-1)^{n+m}$$
$$K_{n+m+1}(m_j s) I_{n+m}(m_j r))\} \tag{101}$$

Thus the total horizontal force components exerted on the cylinder 1 due to the presence of cylinder 2, may be written as

$$F_x = F_x^{(1)} + F_{wx}^{(2)} + F_{dx}^{(2)} + F_{qx}^{(2)} \tag{102}$$
$$F_y = F_y^{(1)} + F_{wy}^{(2)} + F_{dy}^{(2)} + F_{qy}^{(2)} \tag{103}$$

# 8 Drift forces on an isolated cylinder

For a diffracted wave the total horizontal force, $F_\ell$, can be obtained by

$$F_\ell = \int_{\theta=0}^{2\pi} \{ \int_{z=-h}^{0} (-\rho \frac{\partial \phi_\ell}{\partial t}) dz \}_{r=a} (-\cos\theta) a \, d\theta. \tag{104}$$

This expression has been obtained by many researchers including MacCamy and Fuchs (1954).

The second order contributions may be computed by estimating $F_d$ and $F_w$. The dynamic force, $F_d$ is given as:

$$F_d = \int_{\theta=0}^{2\pi} \{ \int_{z=-h}^{0} -\frac{1}{2}\rho(\nabla\phi_\ell)^2 dz \} (-\cos\theta) a \, d\theta. \tag{105}$$

Invoking the expression for $\phi_\ell$ and carrying out the $z$-integration of the expression $(\nabla\phi_\ell)^2$, we obtain (see Rahman (1984)):

$$\begin{aligned}
&-\frac{A^2 g}{8} \sum_{m=0}^{\infty} \sum_{n=0}^{\infty} \delta_m \delta_n R_m R_n \frac{mn}{k^2 a^2} (1 + \frac{2kh}{\sinh 2kh}) \\
&\times \cos\{(m+n)\frac{\pi}{2} + \beta_m + \beta_n - 2\sigma t\} \\
&+ \cos\{(m-n)\frac{\pi}{2} + \beta_m - \beta_n\} [\cos(m+n)\theta - \cos(m-n)\theta] \\
&+ \frac{A^2 g}{8} \sum_{m=0}^{\infty} \sum_{n=0}^{\infty} \delta_m \delta_n R_m R_n (1 - \frac{2kh}{\sinh 2kh}) \\
&\times [\cos\{(m+n)\frac{\pi}{2} + \beta_m + \beta_n - 2\sigma t\} \\
&+ \cos\{(m-n)\frac{\pi}{2} + \beta_m - \beta_n\}] \\
&\times [\cos(m+n)\theta + \cos(m-n)\theta]
\end{aligned} \tag{106}$$

where the Wronskian property of Bessel functions gives:

$$A_m(ka) = \frac{-2i}{(\pi ka) H_m^{(2)\prime}(ka)} = R_m e^{-i\beta_m} \tag{107}$$

in which

$$R_m = (\frac{2}{\pi ka})\{J_m'^2(ka) + Y_m'^2(ka)\}^{-1/2} \quad (108)$$

$$\beta_m = \tan^{-1}[\frac{J_m'(ka)}{Y_m'(ka)}]. \quad (109)$$

In view of the subsequent $\theta$-integration described in Equation (105), we require only the coefficient of $\cos\theta$ in the double summation in (106), and this yields the following:

$$\frac{4A^2 g}{\pi^2 k^2 a^2}\sum_{\ell=0}^{\infty}[(1-\frac{2kh}{\sinh 2kh}) + \frac{\ell(\ell+1)}{k^2a^2}(1+\frac{2kh}{\sinh 2kh})] \times[E_\ell - (-1)^\ell\{C_\ell \cos 2\sigma t - S_\ell \sin 2\sigma t\}] \quad (110)$$

where

$$E_\ell = (J_\ell' Y_{\ell+1}' - J_{\ell+1}' Y_\ell')/T_\ell$$
$$C_\ell = (Y_\ell' J_{\ell+1}' - Y_{\ell+1}' J_\ell')/T_\ell$$
$$S_\ell = (Y_\ell' Y_{\ell+1}' - J_\ell' J_{\ell+1}')/T_\ell$$
$$T_\ell = (J_\ell'^2 + Y_\ell'^2)(J_{\ell+1}'^2 + Y_{\ell+1}'^2)$$

and the Bessel functions arguments are $ka$. Integrating with respect to $\theta$, the total dynamic force, $F_d$, may be expressed as:

$$F_d = \frac{2\rho g A^2}{\pi a k^2}\sum_{\ell=0}^{\infty}[(1-\frac{2kh}{\sinh 2kh}) + \frac{\ell(\ell+1)}{k^2a^2}(1+\frac{2kh}{\sinh 2kh})] \times [E_\ell - (-1)^\ell\{C_\ell \cos 2\sigma t - S_\ell \sin 2\sigma t\}] \quad (111)$$

The result (111) is the sum of the steady-state and oscillatory parts which can be written as:

$$F_d^{ss} = \frac{2\rho g A^2}{\pi a k^2}\sum_{\ell=0}^{\infty}[(1-\frac{2kh}{\sinh 2kh}) + \frac{\ell(\ell+1)}{k^2a^2} \times (1+\frac{2kh}{\sinh 2kh})]E_\ell \quad (112)$$

$$F_d^{os} = -\frac{2\rho g A^2}{\pi a k^2}\sum_{\ell=0}^{\infty}(-1)^\ell\{(1-\frac{2kh}{\sinh 2kh}) + \frac{\ell(\ell+1)}{k^2a^2} \times (1+\frac{2kh}{\sinh 2kh})\}\{C_\ell \cos 2\sigma t - S_\ell \sin 2\sigma t\} \quad (113)$$

such that

$$F_d = F_d^{ss} + F_d^{os}. \quad (114)$$

The waterline force, $F_w$, may be calculated from

$$F_w = \int_0^{2\pi} \frac{\rho}{2g}(\frac{\partial \phi_\ell}{\partial t})^2_{\substack{z=0\\ r=a}}(-\cos\theta)a\,d\theta. \quad (115)$$

Now we know that

$$
\begin{aligned}
(\frac{\partial\phi}{\partial t})^2_{z=0,r=a} &= -\frac{A^2g^2}{4}\sum_{m=0}^{\infty}\sum_{n=0}^{\infty}\delta_m\delta_n R_m R_n[\cos((m+n)\frac{\pi}{2} \\
&\quad +\beta_m+\beta_n-2\sigma t)-\cos((m-n)\frac{\pi}{2}+\beta_m-\beta_n)] \\
&\quad \times[\cos(m+n)\theta+\cos(m-n)\theta]. \qquad (116)
\end{aligned}
$$

In view of the subsequent $\theta$-integration given in (115) we require only the coefficient of $\cos\theta$ term in the double summation in (116) and this yields the following:

$$
\frac{8A^2g^2}{\pi^2k^2a^2}\sum_{\ell=0}^{\infty}[E_\ell+(-1)^\ell\{C_\ell\cos 2\sigma t-S_\ell\sin 2\sigma t\}] \qquad (117)
$$

where $E_\ell$, $C_\ell$, and $S_\ell$ are as previously defined.

Integrating with respect to $\theta$, the total waterline force, $F_\omega$, may be obtained as:

$$
F_\omega=-\frac{4\rho gA^2}{\pi ak^2}\sum_{\ell=0}^{\infty}[E_\ell+(-1)^\ell\{C_\ell\cos 2\sigma t-S_\ell\sin 2\sigma t\}] \qquad (118)
$$

This result may be expressed as the steady-state and oscillatory parts as follows:

$$
F_\omega=F_\omega^{ss}+F_\omega^{os} \qquad (119)
$$

where

$$
F_\omega^{ss}=-\frac{4\rho gA^2}{\pi ak^2}\sum_{\ell=0}^{\infty}E_\ell \qquad (120)
$$

$$
F_\omega^{os}=-\frac{4\rho gA^2}{\pi ak^2}\sum_{\ell=0}^{\infty}(-1)^\ell\{C_\ell\cos 2\sigma t-S_\ell\sin 2\sigma t\}. \qquad (121)
$$

Combining the steady-state and oscillatory parts of $F_d$ and $F_\omega$, we obtain:

$$
F_{d\omega}^{ss} = -\frac{2\rho gA^2}{\pi ak^2}\sum_{\ell=0}^{\infty}[(1-\frac{\ell(\ell+1)}{k^2a^2})(1+\frac{2kh}{\sinh 2kh})E_\ell] \qquad (122)
$$

$$
\begin{aligned}
F_{d\omega}^{0s} &= -\frac{2\rho gA^2}{\pi ak^2}\sum_{\ell=0}^{\infty}(-1)^\ell[(3-\frac{2kh}{\sinh 2kh})+\frac{\ell(\ell+1)}{k^2a^2} \\
&\quad \times\ (1+\frac{2kh}{\sinh 2kh})]\times[C_\ell\cos 2\sigma t-S_\ell\sin 2\sigma t]. \qquad (123)
\end{aligned}
$$

The drift forces for a single cylinder can then be obtained as

$$
F=F_\omega^{ss}+F_d^{ss}. \qquad (124)
$$

We know that the drift forces for a pair of cylinders are

$$
\begin{aligned}
F_x &= (F_{\omega x}^{(2)})_{ss}+(F_{dx}^{(2)})_{ss} \qquad (125) \\
F_y &= (F_{\omega y}^{(2)})_{ss}+(F_{dy}^{(2)})_{ss}. \qquad (126)
\end{aligned}
$$

Therefore the amplification factor for two cylinders as compared to a single cylinder can be estimated by using the following definitions $\mu_x = \frac{F_x}{F}$ and $\mu_y = \frac{F_y}{F}$. These are computed in Table 1 and compared with $\lambda_x$ and $\lambda_y$ (amplification factors for first order forces ) for a variety of parameter values. We also redefine the following dimensionless second order waterline and dynamic force components respectively:

The waterline force components:

$$F_{wx} = \frac{F_{wx}^{(2)}}{2\rho g A a^2} \tag{127}$$

$$F_{wy} = \frac{F_{wy}^{(2)}}{2\rho g A a^2} \tag{128}$$

The dynamic force components:

$$F_{dx} = \frac{F_{dx}^{(2)}}{2\rho g A a^2} \tag{129}$$

$$F_{dy} = \frac{F_{dy}^{(2)}}{2\rho g A a^2} \tag{130}$$

Table 1: Numerical results of amplification factors for various parameters $s/a = 5$, $h/a = 3$

| $ka$ | $ks$ | $kh$ | $\mu_x$ | $\mu_y$ | $\lambda_x$ | $\lambda_y$ |
|---|---|---|---|---|---|---|
| 0.50 | 2.50 | 1.50 | 1.296572 | 0.020009 | 0.965637 | 0.110587 |
| 0.60 | 3.00 | 1.80 | 1.039462 | -0.098980 | 0.939990 | 0.129585 |
| 0.70 | 3.50 | 2.10 | 0.870259 | -0.1235236 | 0.933213 | 0.136623 |
| .80 | 4.00 | 2.40 | 0.783540 | -0.106307 | 0.945425 | 0.141666 |
| 0.90 | 4.50 | 2.70 | 0.752349 | -0.065245 | 0.970223 | 0.150065 |
| 1.00 | 5.00 | 3.00 | 0.755871 | -0.001490 | 0.997097 | 0.163120 |
| 1.10 | 5.50 | 3.30 | 0.781834 | 0.078732 | 1.016313 | 0.180453 |
| 1.20 | 6.00 | 3.60 | 0.833345 | 0.157476 | 1.023968 | 0.200459 |
| 1.30 | 6.50 | 3.90 | 0.924884 | 0.208889 | 1.022195 | 0.220331 |
| 1.40 | 7.00 | 4.20 | 1.053781 | 0.210658 | 1.015537 | 0.237171 |
| 1.50 | 7.50 | 4.50 | 1.181923 | 0.158663 | 1.008307 | 0.250156 |
| 1.60 | 8.00 | 4.80 | 1.264259 | 0.067411 | 1.004103 | 0.261446 |
| 1.70 | 8.50 | 5.10 | 1.282858 | -0.047339 | 1.005507 | 0.274717 |
| 1.80 | 9.00 | 5.40 | 1.241272 | -0.180359 | 1.012547 | 0.293005 |
| 1.90 | 9.50 | 5.70 | 1.143017 | -0.327039 | 1.021617 | 0.316342 |
| 2.00 | 10.00 | 6.00 | 0.995994 | -0.452463 | 1.027404 | 0.337502 |

# 9 Numerical results and conclusion

Numerical computations were carried out to provide results for different parameters for second order forces on a single cylinder. The real parts of the dimensionless components $f_m^{(2)}$, $f_0^{(2)}$, $f_{qII}^{(2)}$ and $f_{qIII}^{(2)}$ are plotted in Figs 2 and 4 for $h/a = 1.0$ and 4.0 respectively as functions of the wave diffraction parameter $ka$.

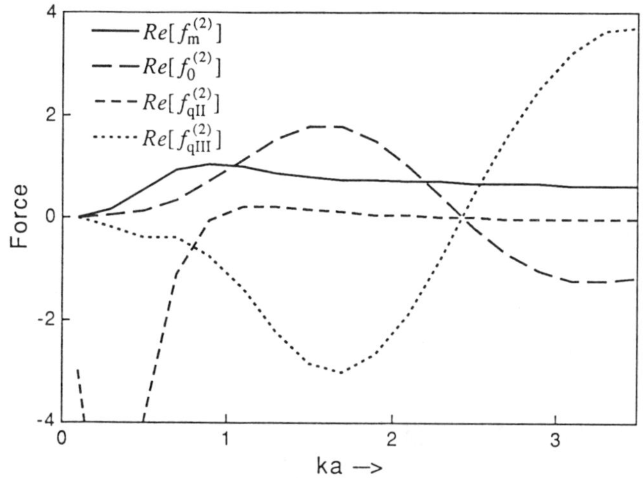


Figure 2: Real parts of the second order dimensionless forces for h/a = 1.0.

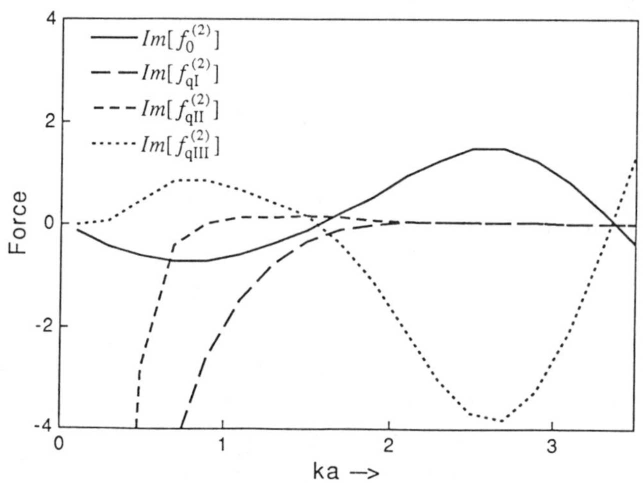


Figure 3: Imaginary parts of the second order dimensionless forces for h/a = 1.0.

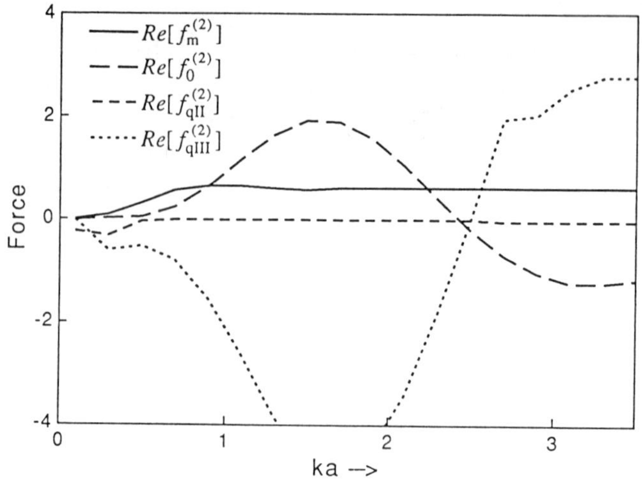


Figure 4: Real parts of the second order dimensionless forces for h/a = 4.0.

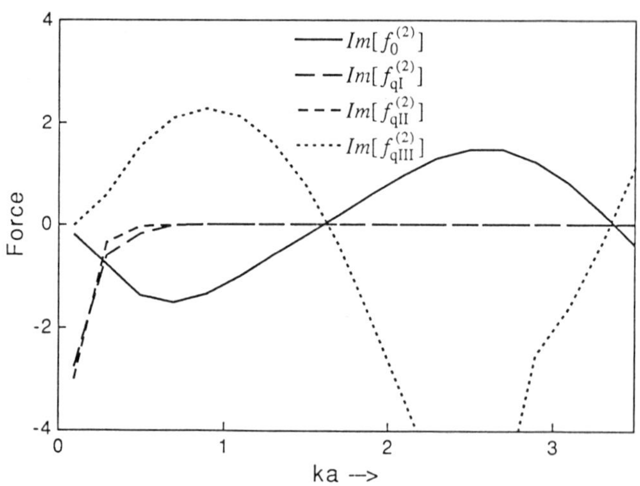


Figure 5: Imaginary parts of the second order dimensionless forces for h/a = 4.0.

Figs 3 and 5 show the imaginary parts of the dimensionless components $f_0^{(2)}$, $f_{qI}^{(2)}$, $f_{qII}^{(2)}$ and $f_{qIII}^{(2)}$ for $h/a = 1.0$ and 4.0 respectively. Double precision arithmetic was used for all the computations on the main frame HP9000/755. IMSL subroutines were used to perform the numerical integration (Gaussian quadrature) and to compute the Bessel functions and the Fresnel sine and cosine functions.

It is seen that $f_0^{(2)}$ and $f_{qIII}^{(2)}$ are 180 degree out of phase in all cases. The real parts of the force $f_m^{(2)}$ and $f_{qII}^{(2)}$ approaches to constants as $ka$ becomes larger. Similar trend is seen for the imaginary parts of $f_{qI}^{(2)}$ and $f_{qII}^{(2)}$ for large $ka$. As $ka$ approaches zero, the imaginary parts become singular and tends to negative infinity.

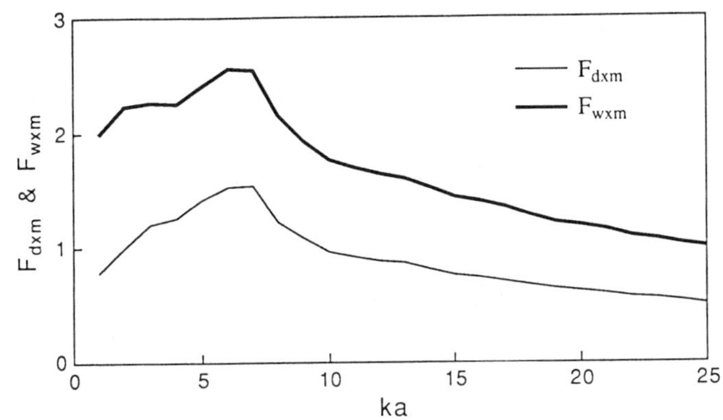


Figure 6: Absolute maximum waterline force $F_{wxm}$ and absolute maximum dynamic force $F_{dxm}$ as functions of ka for kh = 30, A/a = 2 and ks = 50.0.

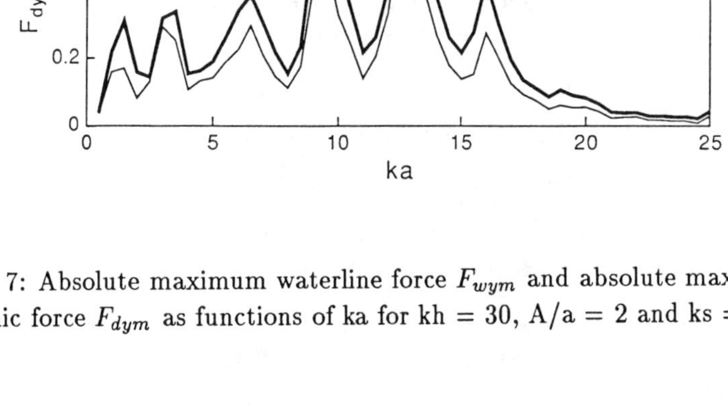


Figure 7: Absolute maximum waterline force $F_{wym}$ and absolute maximum dynamic force $F_{dym}$ as functions of ka for kh = 30, A/a = 2 and ks = 50.0.

The variations in $F_{dxm}$ and $F_{wxm}$ with $ka = 4$ for $kh = 30$ and $A/a = 2$ are shown in Figure 6. Here we take $ks = 50$. It is seen that not only both components are qualitatively similar in their variation with respect to $ka$, but also they are similar in magnitude. $F_{dxm}$ and $F_{wxm}$ attain maxima at about $ka = 7$ and monotonically decrease thereafter. The variation in $F_{dym}$ and $F_{wym}$, shown in Figure 7, is more complicated. In this case we obtain a family of resonance peaks.

## Acknowledgement

The authors are thankful to Natural Sciences and Engineering Research Council of Canada for financial support leading to this work.

## References

1. Abramowitz, M. & Stegun, I. (1972) *Handbook of Mathematical Functions*, National Bureau of Standards, Dover Publications, Inc., New York.

2. Chakrabarti, S.K. (1978) Wave forces on multiple vertical cylinders, *J. Waterways, Port, Coastal and Ocean Division*, ASCE 104 (WW2), 147-154.

3. Eatock Taylor, R. & Hung, S.M. (1985) Wave drift enhancement effects in multicolumn structures, *Applied Ocean Research*, 7(3), 128-139.

4. Eatock Taylor, R. & Hung, S.M. (1987) Second order diffraction forces on a vertical cylinder in regular waves, *Applied Ocean Research*, 9(1), 19-30.

5. Lighthill, M.J. (1979) Waves and hydrodynamic loading, *Behaviour of Offshore Structures Conf.*, Boss'79, Imperial College, London, U.K., pp. 1-40.

6. MacCamy, R.C. & Fuchs, R.A. (1954) Wave force on piles: A diffraction theory, *Techn. Memo No. 69*, Beach Erosion Board, Coastal Engineering Research Center, U.S. Army, Washington, D.C.

7. Masuda, K., Kato, W. & Ishizuka, H. (1986) Second order diffraction loads on phase vertical cylinder with arbitrary cross sections, *Proc. 5th Intl. OMAE Symposium*, Tokyo, Japan, pp. 345-352.

8. Masuda, K., Kato, W. & Sakata, C. (1987) On linear effects of surface waves on wave loads upon the plural vertical cylinder with arbitrary cross sections, *Proc. 6th Intl. OMAE Symposium*, Houston, U.S.A., pp. 72-81.

9. McIver, P. & Evans, D.V. (1984) Approximation of wave forces on cylinder arrays, *Applied Ocean Research*, Vol. 6, 101-107.

10. Molin, B. (1979) Second order diffraction loads on three-dimensional bodies, *Applied Ocean Research*, 1(4), 197-202.

11. Ohkusu, M. (1973) Wave actions on groups of vertical circular cylinders. *In selected papers from J. Soc. Nav. Arch.*, Japan, Vol. 11, 37-50.

12. Rahman, M. (1984) Wave diffraction by large offshore structures: An exact second order theory, *Applied Ocean Research*, 6, 90-100.

13. Rahman, M. (1987) A design method of predicting second order wave diffraction caused by large offshore structures, *Ocean Engg.*, 14(1), 1-8.

14. Simon, M.J. (1982) Multiple scattering in arrays of axisymmetric wave energy devices, Part 1: A matrix method using a plane wave approximation, *J. Fluid Mechanics*, Vol. 18, 273-285.

15. Spring, B.H. & Monkmeyer, P.L. (1974) Interaction of plane wave with vertical cylinders, *Proc. 14th Int. Conf. on Coastal Engineering*, Copenhagen, Denmark, ASCE, pp. 1828-1847.

# Chapter 6

# Recent advances in numerical modelling of non-linear propagation of water waves and their interaction with impermeable bodies

**J.S. Pawlowski, R.E. Baddour**

*National Research Council of Canada, Institute for Marine Dynamics, Computational Hydrodynamics, St. John's, NF, Canada, A1B 3T5*

**Abstract**

The modelling of hydrodynamic problems which involve an unknown free surface with non-linear and time dependent boundary conditions is discussed in the light of a novel domain mapping technique. The numerical techniques found to be suitable and reliable in obtaining numerical solutions to the steady and unsteady non-linear boundary value problems are reviewed. Specific applications of the presented theoretical and numerical modelling techniques are described. Results of computations for selected problems are presented.

## 1 Introduction

The inherent difficulty in the modelling of a class of hydrodynamic problems consists in the presence of the unknown free surface on which non-linear and time dependent boundary conditions are applied. At present, because of their level of difficulty, those problems are primarily dealt with using the assumption of potential flow. In this chapter a discussion of the general problem of the non-linear interaction of surface waves in ideal fluid with fixed or moving impermeable bodies is presented.

The second and third sections of this chapter introduce and provide examples of the use of a novel analytic mapping method developed by the first author. This method makes posssible consistent perturbation formulations and solutions of non-linear boundary value problems, including consistent construction of a non-linear wave radiation condition. It also facilitates direct formulations and solutions of non-linear boundary value problems. The examples include novel formulations of steady and unsteady non-linear boundary value problems.

The fourth section of this chapter describes the boundary element semi-discretization techniques which have been found by the authors to be suitable and reliable in obtaining numerical solutions to the non-linear steady and transient boundary value problems. The last section describes specific applications of the presented theoretical and numerical modelling techniques. The applications are based on the example formulations described earlier and comprise a novel non-linear, time domain, numerical model of ship interaction with heavy seas, including the use of the weak-scatterer hypothesis, and a novel non-linear model of the generation of waves and wave resistance by a ship advancing in calm water. The last section uses the results obtained by the authors and their collaborators over the last five years.

The references provided in the text correspond to the need of leading the reader to additional specific information or to other references which provide background information. No attempt is made to give a comprehensive literature review.

# 2 Fundamentals of the Analytic Mapping Method

## 2.1 The Definition and Properties of Analytic Mappings

Analytic mappings defined and discussed below are mappings of domains in a three-dimensional Euclidean space $E$ and of tensor fields defined on those domains. The concept of a domain mapping is a simple one. Let $D_0$ denote a time dependent domain (an open connected set) in $E$. A relation :

$$y(x,t) = x + \eta(x,t), \quad x \in D_0,\, y \in D \tag{1}$$

with $x$ in $D_0$, $y$ in $D$, where $D$ is another domain in $E$, and $t$ denoting the time variable, defines a mapping of $D_0$ onto $D$. The mapping is assumed to be one-to-one (bijective), and continuous together with its inverse. Such a mapping is a homeomorphism between $D_0$ and $D$. The mapping transforms in a one-to-one manner open sets onto open sets, closed sets onto closed sets, and boundaries of sets onto boundaries of their images under the transformation. In the sequel the homeomorphism will be called the displacement mapping, by analogy to the Lagrangian description of motion. By the same analogy $\eta(x,t)$ will be called the displacement field.

Assuming that $\eta(x,t)$ is chosen so that mapping (1) is bijective, in order to obtain a displacement mapping it is sufficient, but not necessary, to require that $\eta(x,t)$ is continuously differentiable with respect to the space variable (radius vector) $x$ and that the norm of the derivative satisfies the inequality:

$$\mid \partial_x \otimes \eta(x,t) \mid < 1. \tag{2}$$

In (2) the symbol of partial differentiation with the subscript $x$ denotes the gradient operator, and $\otimes$ signifies tensor multiplication.

It can be observed that mapping (1) is analytic on $D$. Indeed it is given by :

$$y = x + (y - x) \cdot I \tag{3}$$

which is a power series relative to $y - x$, with $I$ denoting the unit tensor. Relation (3) can be rewritten in the form of a Taylor's series:

$$y = x + \eta \cdot \partial_x \otimes x. \tag{4}$$

It is important to notice that equations (3) and (4) can be considered as a transformation of radius vector field $x$ on $D_0$ onto radius vector field $y$ on $D$. This transformation is generalized to other tensor fields of interest on $D$ and $D_0$, (Pawlowski, 1992) and (Pawlowski, 1994).

In order to carry out the required generalizations, the tensor fields are assumed to be represented by analytic functions on sufficiently large open intervals in $E$. An open interval $L$ in $E$ is defined by the following inequalities:

$$0 \leq |(y - x)_i| < r_i, \quad i = 1, 2, 3 \tag{5}$$

satisfied by the three components of the radius vector $y - x$. In (5) $r_i$ denotes any positive number of the extended real number axis which includes $\infty$. A real valued function $f(y)$ which is represented on an open interval $L$ by an absolutely convergent power series :

$$f(y) = \sum_{\nu=0}^{\infty} a_{ijk}(y - x)_1^i (y - x)_2^j (y - x)_3^k \tag{6}$$

with $\nu = i + j + k$, is analytic on that interval. This series is unique and identical with the Taylor's series:

$$f(y) = \sum_{\nu=0}^{\infty} \frac{1}{\nu!} [(y - x) \cdot \partial_x]^\nu f(x). \tag{7}$$

In (7) the following notation is used :

$$[(y - x) \cdot \partial_x]^\nu = \nu! \sum_{i+j+k=\nu} \frac{1}{i!j!k!} (y - x)_1^i (y - x)_2^j (y - x)_3^k \, \partial_{x_1}^i \partial_{x_2}^j \partial_{x_3}^k. \tag{8}$$

The above expression is equal to 1 for $\nu = 0$. Formula (7) can be rewritten in a convenient exponential operator form:

$$f(y) \doteq \exp[(y - x) \cdot \partial_x] \, f(x) \tag{9}$$

which corresponds to the notation used in the sequel:

$$\exp(a \cdot \partial_x) \doteq \sum_{\nu=0}^{\infty} \frac{1}{\nu!} (a \cdot \partial_x)^\nu \tag{10}$$

for a vector $a$. As an example equation (4) is rewritten in the form:

$$y = \exp(\eta \cdot \partial_x) \, x. \tag{11}$$

It should be understood that in general $y$ and $x$ cannot be exchanged for each other in (9) unless $f(x)$ is an entire function (which is the case in (11)).

The series representation of an analytic function is generalized if one considers an analytic function $F(u)$ of $k$ real arguments $u_i$, defined by the series:

$$F(u) = \exp[\sum_{i=1}^{k}(u_i - v_i)\partial_{v_i}]\, F(v) \tag{12}$$

in an interval:

$$0 \leq |(u_i - v_i)| < p_i \tag{13}$$

for $i = 1$ to $k$. In addition, the components $u_i$ of $u$ are assumed to be analytic functions defined in an open interval in $D$ by:

$$u_i(y) = \exp[(y - x) \cdot \partial_x]\, v_i(x). \tag{14}$$

If $y$ is chosen so that $u_i(y)$ satisfy conditions (13) then series (14) can be used in series (12) to obtain:

$$F(u) = \exp[(y - x) \cdot \partial_x]\, F(v). \tag{15}$$

Finally, we consider two analytic functions $f_i(z)$, $i = 1, 2$, defined in the open interval $L$ (see (5) above):

$$f_i(z) = \exp[(z - x) \cdot \partial_x]\, f_i(x). \tag{16}$$

Let another function $f(y)$ be given in an open interval contained in $L$, and such that:

$$f(y) = \exp[(y - z) \cdot \partial_z]\, f_i(z) \tag{17}$$

for $i = 1, 2$. From (16) and (17) it is found that

$$f(y) = \exp[(y - x) \cdot \partial_x]\, f(x) \tag{18}$$

where $f(x) = f_1(x) = f_2(x)$ on $L$. The above brief review of relevant properties of analytic functions, using (Dieudonné, 1969), assists the discussion of the analytic mapping method as applied to general tensor fields.

Let $D$ represent a fluid domain in $E$. The sub-domains of $D$ are a finite number of disjoints domains $D_i$ in $E$ such that:

$$\overline{D} = \bigcup_i \overline{D}_i \tag{19}$$

where an over-line denotes the closure of a set. At any moment in time $t$, the sub-domains $D_i$ are defined as images of the corresponding reference sub-domains $D_{0_i}$ of a reference fluid domain $D_0$:

$$\overline{D}_0 = \bigcup_i \overline{D}_{0_i}. \tag{20}$$

The mapping from the reference domain onto the instantaneous domain is a displacement mapping defined by (1).

For all values of $t$, the closures of $D_{0_i}$ and $D_i$ are assumed to remain in an open interval $L_i$. In addition the scalar components $u_i$ of any tensor field $u$, such as a fluid velocity field, that describes a fluid flow on $D_i$, are assumed to be represented in $L_i$ by absolutely convergent power series:

$$u_j(y) = \exp\{[y - x(i)] \cdot \partial_x\} \, u_j[x(i)] \tag{21}$$

where $x(i)$ is the midpoint of $L_i$. It follows from (17) and (18) that if:

$$u_j(y) = \exp(\eta \cdot \partial_x) \, v_j(x) \tag{22}$$

for $y$ in the closure of $D_i$ and $x$ in the closure of $D_{0_i}$, then $v_j(x)$ is independent of the chosen displacement field $\eta$, and $v_j(x) = u_j(x)$ on $L_i$.

Formula (21) says that the scalar components of tensor field $u$ are represented on the instantaneous fluid sub-domain by functions which, in an appropriately chosen interval in $E$, are expressed by absolutely convergent power series of the radius vector relative to the midpoint of the interval. This assumption is not too restrictive since every function continuous together with its first $k$ derivatives on a compact domain in $E$ can be approximated on that domain arbitrarily closely, together with the derivatives, by a polynomial (Birkhoff and Lynch, 1984). The assumption is always satisfied by entire functions.

Formula (22) determines an analytic mapping of the tensor field $v(x)$ defined on the closure of $D_{0i}$, to the tensor field $u(y)$ defined on the closure of $D_i$. In this respect formula (22) is completely analogous to formulae (4) and (11). The analytic mapping is constructed using the fact that the instantaneous sub-domain is contained in the interval $L_i$ which is the domain of the tensor field defined by (21). On the closure of $D_i$ this tensor field is identical with the field $u(y)$. The displacement field $\eta$ must be such that the series on the right-hand side of (22) converges. If the components of field $u(y)$ are entire functions the latter condition does not impose any limitation on displacement field $\eta$ .

The tensor fields on sub-domains $D_i$ must satisfy appropriate governing equations (i.e. domain equations and boundary conditions) pertinent to the boundary value problem which governs a flow problem under consideration. The division of the fluid domain $D$ into sub-domains $D_i$ allows one to deal with possible flow discontinuities. Such discontinuities can be imposed on the boundaries between the sub-domains through the boundary conditions. Therefore, in general, the solutions to the boundary value problems are sought in a piecewise analytic form. It can be observed that the practice of using piecewise polynomial approximations in numerical solutions to boundary value problems agrees with the described piecewise analytic formulation. In the examples discussed below the subdivision of the domain will not be used unless it is explicitly referred to.

The purpose of introducing the analytic mappings of domains and tensor fields is to be able to consistently represent non-linear and transient boundary value problems of fluid dynamics by governing equations formulated on fixed in space, time independent and conveniently chosen reference domains. Such representations ensure consistent and systematic development of formulations and solutions to the non-linear and transient problems, including non-linear wave radiation conditions, and significantly increase the possibility of constructing accurate and efficient algorithmic solutions.

A different mapping method was considered in (Lebovitz, 1982) as a justification of the usual perturbation formulations of boundary value problems. The procedure proposed by Lebovitz uses the identity mapping $u(y) = v(x)$, in the present notation, and expands $u(y)$ analytically relative to $\eta$ in a perturbation series. In comparison the method presented in (Pawlowski, 1992), (Pawlowski, 1994) and in this discussion, defines the domain and tensor field mappings consistently and avoids the mixing of the analytic mapping with a perturbation scheme. As a result, the analytic mapping method presently discussed is clearly defined and capable of treating such problems as the formulation of explicit impermeability conditions and the analytical requirements imposed on the mapped fields.

Having defined the analytic mappings, the next step is to use them to transform the governing equations defined on an instantaneous fluid domain into governing equations defined on a reference domain.

## 2.2 Transformations of The Governing Equations

Governing equations of fluid dynamics are expressed in terms of derivatives and integrals of components of tensor fields, such as fluid velocity and pressure fields (the discussion below is restricted to examples of ideal fluid flow), which describe fluid flow on an instantaneous fluid domain. Therefore it is necessary to determine the manner in which these derivatives and integrals can be expressed by means of derivatives and integrals of components of the corresponding tensor fields defined on a reference fluid domain. Once the transformations of the derivatives and integrals of the tensor fields are established, the governing equations can be transformed to represent the boundary value problems on the reference fluid domain. The described above homeomorphic equivalence of the domains results in a consistent transformation of the governing equations. This means that equations of a given type which apply on a part or boundary (such as, e.g. an impermeable boundary) of the instantaneous domain are transformed in a unique manner to equations applicable on the corresponding part or boundary of the reference domain.

Using equations (1) and (22) it can be shown (Pawlowski, 1992) that the mapping formula (22) applies to spatial derivatives of the tensor fields, of arbitrary order, and to time derivatives of the tensor fields, when they exist:

$$\partial^k_{y_1}\partial^l_{y_2}\partial^m_{y_3}\partial^n_t\, u_i(y) = \exp(\eta\cdot\partial_x)\,\partial^k_{x_1}\partial^l_{x_2}\partial^m_{x_3}\partial^n_t\, v_i(x). \tag{23}$$

In addition, if $F(u)$ is an analytic function of components $u_i$ of tensor field $u(y)$ then, under conditions discussed in connection with formulae (12)-(15), it follows from (14), (15) and (22) that:

$$F(u) = \exp(\eta\cdot\partial_x)\,F(v). \tag{24}$$

The consideration of conditions applied on the domain boundaries, such as the impermeability condition, requires the knowledge of the mapping of surfaces. The defining equation of a surface in the instantaneous domain, which is the image under a displacement mapping of a surface determined by the formula:

$$F(x) = C \tag{25}$$

is written in the form :

$$f(y,t) - F(x) = 0. \tag{26}$$

Following (Pawlowski, 1992) formulae (26) and (11) result in:

$$\partial_y f(y,t) = (I + \partial_x\otimes\eta)^{-1}\cdot\partial_x F(x) \tag{27}$$

and

$$\partial_t f(y,t) = -\partial_t\eta\cdot(I+\partial_x\otimes\eta)^{-1}\cdot\partial_x F(x). \tag{28}$$

In (27) and (28) the superscript $-1$ denotes the inverse of the tensor in brackets. It should be observed that under condition (2) this inverse is given by the Neumann series:

$$K \doteq (I+\partial_x\otimes\eta)^{-1} = I - \partial_x\otimes\eta + (\partial_x\otimes\eta)\cdot(\partial_x\otimes\eta) + \cdots \tag{29}$$

A closed form of $K$ can be derived (Pawlowski, 1992) :

$$K = \frac{1}{1+\sum_{i=1}^3 J_i}[(1+\partial_x\cdot\eta)I - \partial_x\otimes\eta + \frac{1}{2}\sum_{i,j,k=1}^{3}\epsilon_{ijk}\,\partial_{x_i}\eta\wedge\partial_{x_j}\eta\otimes e_k]. \tag{30}$$

In (30) $J_i$, $i = 1,2,3$, denote the scalar invariants of the gradient tensor of $\eta$, $\epsilon_{ijk}$ is the sign of index permutation, and $e_k$, $k = 1,2,3$, represent the unit vectors of a Cartesian system of reference in which the formula is expressed. Formula (27) results in the following relation between the normal vector $n$ to the surface in the instantaneous fluid domain and the normal vector $N$ to the surface in the reference domain at the corresponding location:

$$n = K \cdot N \frac{1}{\sqrt{N \cdot K^T \cdot K \cdot N}} \tag{31}$$

with superscript $T$ denoting the transpose of $K$.

When the surface defined by (25) is parametrized so that:

$$F[x(\lambda, \mu)] = C \tag{32}$$

where $\lambda$ and $\mu$ are real parameters, the differential element of the surface is defined by:

$$dS_x \doteq \partial x_\lambda \wedge \partial x_\mu \tag{33}$$

with:

$$\partial x_\lambda \doteq \partial_\lambda x \, d\lambda \tag{34}$$

and:

$$\partial x_\mu \doteq \partial_\mu x \, d\mu. \tag{35}$$

The corresponding differential element on the instantaneous surface is given by the formula :

$$dS_y = \partial y_\lambda \wedge \partial y_\mu. \tag{36}$$

Using the above adopted notation one obtains:

$$dS_y = (1 + \sum_{i=1}^{3} J_i) K \cdot dS_x. \tag{37}$$

In an analogous manner corresponding differential volume elements $dV_y$ and $dV_x$ can be considered using three space parameters $\lambda$, $\mu$ and $\gamma$. Then :

$$dV_x \doteq \partial x_\gamma \cdot (\partial x_\lambda \wedge \partial x_\mu) \tag{38}$$

and:

$$dV_y \doteq \partial y_\gamma \cdot (\partial y_\lambda \wedge \partial y_\mu). \tag{39}$$

with:

$$dV_y = (1 + \sum_{i=1}^{3} J_i) \, dV_x. \tag{40}$$

Although the method of analytic mappings can be applied to the formulation of boundary value problems in general, examples of its application which are discussed below are limited to instances of irrotational ideal fluid flow. Therefore the tensor fields of primary interest are those of velocity potential $\Psi(y, t)$, fluid flow velocity

$u(y,t)$, and fluid pressure $q(x,t)$, all defined on the instantaneous fluid domain $D$. The well-known relations between these fields are expressed as:

$$u(y,t) = \partial_y \Psi(y,t), \quad y \in D \tag{41}$$

and:

$$q(y,t) = -\rho(\partial_t \Psi + \frac{1}{2}|\partial_y \Psi|^2 + g y_3), \quad y \in D \tag{42}$$

with $\rho$ denoting fluid density, $g$ the acceleration of gravity. The constant in the Bernoulli's equation is set to 0, and $y_3$ signifies a vertical Cartesian coordinate with the positive direction taken upwards.

Using the method of analytic mappings, the velocity potential $\Psi(y,t)$ is considered to be the image of a scalar field $\Phi(x,t)$ defined on a reference fluid domain $D_0$. According to formula (22):

$$\Psi(y,t) = \exp(\eta \cdot \partial_x)\, \Phi(x,t). \tag{43}$$

Applying formulae (23) and (24) the following expressions are obtained:

$$u(y,t) = \exp(\eta \cdot \partial_x)\, v(x,t) \tag{44}$$

and:

$$q(y,t) = \exp(\eta \cdot \partial_x)\, p(x,t) \tag{45}$$

where $v(x,t)$ and $p(x,t)$ are defined as:

$$v(x,t) \doteq \partial_x \Phi(x,t) \tag{46}$$

and:

$$p(x,t) \doteq -\rho(\partial_t \Phi + \frac{1}{2}|\partial_x \Phi|^2 + g x_3). \tag{47}$$

In other words velocity field $u(y,t)$ and pressure field $q(x,t)$ defined on the instantaneous fluid domain $D$, are images of the corresponding fields $v(x,t)$ and $p(x,t)$ defined on the reference fluid domain $D_0$. The formulae show that the fields $v(x,t)$ and $p(x,t)$ can be interpreted correspondingly as the velocity and pressure fields on the reference domain, with the field $\Phi(x,t)$ considered as the velocity potential.

The field governing equation for velocity potential $\Psi(y,t)$ in the instantaneous fluid domain is the Laplace's equation:

$$\partial_y^2 \Psi(y,t) = 0, \quad y \in D. \tag{48}$$

In addition on an impermeable boundary, either a wetted surface $\partial D_W$ or free surface $\partial D_F$:

$$(\partial_t + \partial_y \Psi \cdot \partial_y)\, f(y,t) = 0, \quad y \in \partial D_W, \partial D_F \tag{49}$$

where $f(y,t) = C$ defines the boundary under consideration. For flows with a free surface the condition of constant (normalized to 0) pressure is imposed:

$$q(y,t) = 0, \quad y \in \partial D_F \tag{50}$$

together with the impermeability condition. Often it is convenient to impose the impermeability condition on a free surface in the equivalent form of the dynamic condition:

$$(\partial_t + \partial_y \Psi \cdot \partial_y)\, q(y,t) = 0, \quad y \in \partial D_F. \tag{51}$$

Finally, on a geometric internal boundary imbedded in the fluid domain $D$, which can be called control boundary $\partial D_C$, the conditions of the continuity of the velocity and pressure fields are applied (discontinuities of these fields can be imposed in a similar manner) :

$$\Psi(y,t) - \Psi_e(y,t) = 0, \quad y \in \partial D_C \tag{52}$$

and:

$$\partial_y f \cdot \partial_y[\Psi(y,t) - \Psi_e(y,t)] = 0, \quad y \in \partial D_C. \tag{53}$$

In formulae (52) and (53) the subscript $e$ indicates values on the side of the boundary identified by the positive direction of the normal.

An application of formula (24) in equations (48), (50), (51) and (52) gives their transformed equivalents on the reference domain. One obtains respectively:

$$\exp(\eta \cdot \partial_x)\, \partial_x^2 \Phi(x,t) = 0, \quad x \in D_0 \tag{54}$$

$$\exp(\eta \cdot \partial_x)(\partial_t \Phi + \frac{1}{2}|\partial_x \Phi|^2 + g x_3) = 0, \quad x \in \partial D_{0F} \tag{55}$$

$$\exp(\eta \cdot \partial_x)(\partial_t^2 \Phi + g \partial_{x_3} \Phi + 2\partial_x \Phi \cdot \partial_x \partial_t \Phi + \partial_x \Phi \cdot \partial_x \otimes \partial_x \Phi \cdot \partial_x \Phi) = 0, \quad x \in \partial D_{0F} \tag{56}$$

$$\exp(\eta \cdot \partial_x)[\Phi(x,t) - \Phi_e(x,t)] = 0, \quad x \in \partial D_{0C} \tag{57}$$

where the boundaries are the corresponding images of the boundaries of the instantaneous fluid domain. Using formulae (23), (25)-(30) and (41), expressions (49) and (53) are transformed respectively to:

$$[\exp(\eta \cdot \partial_x)\, \partial_x \Phi - \partial_t \eta] \cdot K \cdot N = 0, \quad x \in \partial D_{0W} \tag{58}$$

and:

$$\exp(\eta \cdot \partial_x)\, \partial_x(\Phi - \Phi_e) \cdot K \cdot N = 0, \quad x \in \partial D_{0C}. \tag{59}$$

Taking into account relation (15) and the principle of analytic continuation, equation (54) is reduced to the Laplace's equation:

$$\partial_x^2 \Phi(x,t) = 0, \quad x \in D_0. \tag{60}$$

In instances when the instantaneous configuration of the free surface can be represented as a single valued wave elevation $\eta_3(x,t)$ defined on the horizontal reference free surface at $x_3 = 0$, the exponential operator in equations (55) and (56) is replaced by:

$$\exp(\eta \cdot \partial_x) = \exp(\eta_3 \partial_{x_3}), \qquad x \in \partial D_{0F}. \tag{61}$$

When $\eta$ and its first spatial derivatives are perpendicular to the normal vector at the reference configuration of the impermeable surface, the impermeability condition (58) is simplified to:

$$[\exp(\eta \cdot \partial_x)\, \partial_x \Phi] \cdot N = 0, \qquad x \in \partial D_{0W}. \tag{62}$$

Similarly, if the first spatial derivatives of $\eta$ are perpendicular to the normal vector at the reference configuration of the control boundary, condition (59) takes the simplified form:

$$[\exp(\eta \cdot \partial_x)\, \partial_x(\Phi - \Phi_e)] \cdot N = 0, \qquad x \in \partial D_{0C}. \tag{63}$$

Equations (54) - (63) give a summary of the transformed governing equations of ideal fluid flow, which are useful in applied problems of hydrodynamics. Taking into account formulae (16)-(18), (22) and (43) solutions to boundary value problems defined by the transformed governing equations do not depend on the choice of a displacement field $\eta$.

In the sequel examples of novel formulations of problems in non-linear hydrodynamics are presented as an illustration of the application of the method of analytic mapping.

# 3 Examples of Applications of The Method of Analytic Mappings

## 3.1 The Explicit Impermeability Condition

The impermeability condition is essential in formulations of boundary value problems which model the interaction of a rigid or deformable body with a surrounding fluid. The method of analytic mappings is instrumental in the derivation of the explicit form of the condition (Pawlowski, 1992), and (Pawlowski, 1994). In turn, the explicit form of the impermeability condition allows one to apply this condition in a consistent manner to problems involving deformable bodies and bodies floating on the free surface.

Condition (58) imposed on the reference wetted surface of an impermeable body corresponds to:

$$(u - \partial_t \eta) \cdot n = 0, \qquad y \in \partial D_W \tag{64}$$

where $n$ is the normal vector on the instantaneous wetted surface. Introducing the velocity of the instantaneous wetted surface along its normal, $u_n$, the condition is rewritten as:

$$u \cdot n = u_n, \qquad y \in \partial D_W. \tag{65}$$

This is the well-known form of the impermeability condition. In comparison with (64), in (58) the dependance on the domain displacement field $\eta$ is made explicit. Therefore (58) is named the explicit impermeability condition (Pawlowski, 1992).

It is important to notice that in (58) displacement field $\eta$ defines a mapping between the instantaneous and the reference wetted surfaces. This mapping property is used in the term:

$$u(y,t) = \exp(\eta \cdot \partial_x)\,\partial_x \Phi(x,t), \qquad y \in \partial D_W,\ x \in \partial D_{0W}. \tag{66}$$

Therefore in order to apply the explicit impermeability condition a displacement field $\eta$ must be defined. Following (Pawlowski, 1994) the construction of appropriate displacement fields will be discussed in three stages: first, for a rigid body moving in an infinite fluid domain; second, for a deformable body moving in an infinite fluid domain; and third, for a deformable body moving on the free surface of an otherwise infinite fluid domain.

For a rigid body in motion, its displacement from a reference configuration can be written as:

$$\eta_B(x,t) = \alpha(t) + [R(t) - I] \cdot (x - x_0), \qquad x \in D_{0B} \tag{67}$$

where $D_{0B}$ denotes the reference configuration of the body, $\alpha$ is a parallel translation field, $R(t)$ is a tensor of rotation, and $x_0$ is the radius vector of the center of rotation in the reference configuration. The displacement field $\eta(x,t)$ in the fluid domain is defined as an extension of the body displacement field:

$$\eta_B(x,t) = \alpha(t) + [R(t) - I] \cdot (x - x_0), \qquad x \in D_0. \tag{68}$$

Taking into account that the norm of $R$ is equal to 1, it is established that displacement mapping (68) determines a homeomorphism between $D$ and $D_0$. Therefore displacement field (68) is used in condition (58), and the following explicit form of the impermeability condition is obtained:

$$[\exp(\eta_B \cdot \partial_x)\,\partial_x \Phi - \partial_t \eta_B] \cdot R \cdot N = 0, \qquad x \in \partial D_{0W}. \tag{69}$$

For a deformable body in motion the body displacement field can be written as:

$$\eta_B(x,t) = \alpha(t) + [R(t) - I] \cdot (x - x_0) + R(t) \cdot Q(x,t), \qquad x \in D_{0B} \tag{70}$$

where $Q(x,t)$ denotes a field of deformation. In this case formula (70) cannot be applied to points in the fluid domain because the deformation field is not defined there. A discussion of possible extensions of the displacement field $\eta_B(x,t)$ to the fluid domain is given in (Pawlowski, 1994). Here it is assumed that it is possible to construct a (homeomorphic) displacement field $\eta(x,t)$ on $D_0$, which on $\partial D_{0W}$ takes the values of the body displacement field $\eta_B(x,t)$. Accordingly, after a simple manipulation, the impermeability condition (58) can be written for the case of a deformable body as:

$$[\exp(\eta_B \cdot \partial_x)\partial_x \Phi - \partial_t \eta_B] \cdot R \cdot (I + \partial_x \otimes Q)^{-1} \cdot N = 0, \quad x \in \partial D_{0W}. \tag{71}$$

The impermeability conditions (69) and (71) are applicable to totally submerged bodies. However, when rewritten in terms of quantities defined on the instantaneous wetted surface $\partial D_W$, they take the general form:

$$[u(y,t) - \partial_t \eta_B] \cdot K_B \cdot N = 0 \quad y \in \partial D_W \tag{72}$$

which applies on the instantaneous wetted surface of an impermeable body either fully submerged or not. The crucial difference consists in that in (72) fluid velocity $u$ is defined on the instantaneous wetted surface of the body, whereas in (69) and (71) the expression:

$$u(y,t) = \exp(\eta_B \cdot \partial_x)\partial_x \Phi(x,t) \tag{73}$$

is used, which is defined only on the reference wetted surafce of the body, $\partial D_{0W}$. If the body remains totally submerged its instantaneous wetted surface, $\partial D_W$, coincides with the instantaneous impermeable surface, $\partial D_B$, and the reference wetted surface, $\partial D_{0W}$, is identical with the reference surface of the body, $\partial D_{0B}$. Therefore, since the displacement fields $\eta_B$, used in (69) and (71), are one-to-one mappings between the reference and instantaneous body surface, they are also one-to-one mappings between the reference and instantaneous wetted surface.

In general, for a body floating at a free surface, with the wetted surface constituting only a part of the body surface, the displacement fields (68) and (70), which give the values of the body displacement on the body surface, do not provide one-to-one mappings between the reference and instantaneous wetted surfaces of the body. As a result expression (73) cannot be used in such cases to formulate explicit impermeability conditions. This difficulty is removed, (Pawlowski, 1992), (Pawlowski, 1994), by defining a displacement field on the fluid domain, $D_0$, in such a way that it is compatible with the displacement on the reference wetted surface. Such a displacement field is given by the formula:

$$\eta(x,t) = \eta_0(x,t) + \eta_B(x_B,t), \quad x \in \partial D_{0W} \tag{74}$$

with:

$$x_B = x + \eta_0(x,t). \tag{75}$$

This form of the displacement field on the wetted surface will be assumed below.

## 3.2 Formulations for Steady Flow, Ship Wave Resistance

For steady flow problems governing equations (54), (55) and (58) on reference fluid domain $D_0$ take respectively the form:

$$\partial_x^2 \Phi = 0, \quad x \in \partial D_0 \tag{76}$$

$$\exp(\eta \cdot \partial_x)(\frac{1}{2} \mid \partial_x \Phi \mid^2 + g x_3) = 0 \quad x \in \partial D_{0F} \tag{77}$$

and:

$$\exp(\eta \cdot \partial_x)\partial_x \Phi \cdot K \cdot N = 0, \quad x \in \partial D_{0W},\ \partial D_{0F} \tag{78}$$

Alternatively the free surface condition:

$$\exp(\eta \cdot \partial_x)(g\partial_{x_3}\Phi + \partial_x \Phi \cdot \partial_x \otimes \partial_x \Phi \cdot \partial_x \Phi) = 0, \quad x \in \partial D_{0F}. \tag{79}$$

can be used.

In the above equations the reference free surface $\partial D_{0F}$ is defined by the plane $x_3 = 0$. In addition infinite water depth is assumed, in the absence of impermeable bodies other than a ship hull advancing with a constant forward speed $U$ in the positive direction of the $x_1$ axis. Considering the resulting boundary value problem in a system of reference moving with the ship, the Bernoulli's equation constant $-\frac{1}{2}U^2$ is added to the pressure expression in equation (77). The formulation of the problem is completed by imposing a causality condition which prevents water surface waves generated by the ship from propagating upstream in the positive $x_1$ direction, and requiring that the fluid remains at rest at infinite depth. The causality condition is usually imposed by using an appropriate Green's function in the linear Neumann-Kelvin formulation, or by applying a one-sided differencing formula in a numerical implementation of the free surface condition (79).

The above formulated steady boundary value problem will be referred to as the wave resistance problem, since its solutions are obtained in order to predict the resistance to the ship forward motion. This resistance is a consequence of the generation, by the ship, of a steady wave pattern on the free surface. The above described wave resistance boundary value problem is clearly non-linear with respect to the unknown velocity potential $\Phi$. However, as a result of the application of the analytic mapping method, it is defined on a known fluid domain (assuming that the trim and sinkage of the ship are fixed). The boundaries of the domain are delineated by the reference wetted surface of the ship, which is determined by the hull geometry, and by the free surface which coincides with the horizontal plane $x_3 = 0$. The simplicity of the free surface on which the non-linear boundary condition (79) must be imposed constitutes a primary advantage in constructing a robust and accurate numerical algorithm for the solution to the problem.

In order to handle the non-linearities involved in the boundary conditions, perturbation formulations of the problem are developed (Pawlowski, 1992), (Pawlowski, 1993). The displacement field $\eta$ and velocity potential $\Phi$ are assumed to be represented by perturbation series:

$$\eta = \sum_{i=1}^{\infty} \eta^{(i)} \tag{80}$$

and:

$$\Phi = \sum_{i=0}^{\infty} \Phi^{(i)} \tag{81}$$

where the growing superscripts indicate decreasing orders of magnitude. It should be noticed that the perturbation scheme defines primarily a domain perturbation and threfore, the $\eta$ perturbation series begins with a small displacement field $\eta^{(1)}$ (i.e. $\eta^{(0)} = 0$). In the physical context of the wave resistance problem, this means that the free surface waves generated by the ship are of small amplitude.

In the adopted system of reference the lowest order solution to the problem can be written as:

$$\Phi^{(0)} = -Ux_1 + \phi^{(0)}. \tag{82}$$

The pressure condition on the free surface, with the adjusted Bernoulli's constant, becomes:

$$\mid \partial_x \Phi^{(0)} \mid^2 - U^2 = O(\Phi^{(1)}), \quad x_3 = 0 \tag{83}$$

where symbol $O(.)$ indicates the order of magnitude of the quantity in brackets.

In addition to the assumption of small wave amplitude, generally used formulations of the wave resistance problem involve implicitly the assumption of small wave slope. In the present formulation the assumptions imposed on the displacement field are considered explicitly in the same manner as assumptions applied to the velocity potential. The assumption of small wave slope is expressed by:

$$\mid \partial_x \otimes \eta \mid = O(\Phi^{(1)}) \tag{84}$$

and results in the following impermeability condition on the free surface:

$$\partial_{x_3} \Phi^{(0)} = 0, \quad x_3 = 0. \tag{85}$$

The impermeability condition on the wetted surface of the hull becomes:

$$N \cdot \partial_x \phi^{(0)} = N_1 U, \quad x \in \partial D_{0W}. \tag{86}$$

As is shown below the assumption of small wave slope leads to the generally used at present Neumann-Kelvin and Dawson ( perturbation of the double-body flow ) formulations of the wave resistance problem.

The Neumann-Kelvin formulation is obtained by assuming in the wetted surface impermeability condition:

$$N_1 U = O(\Phi^{(1)}). \tag{87}$$

Then:

$$\Phi^{(0)} = -U x_1 \tag{88}$$

is the highest order of magnitude solution to the problem. For the solution of the next order of magnitude, $\Phi^{(1)}$, the pressure condition on the free surface gives:

$$\eta_3^{(1)} = \frac{1}{g} U \partial_{x_1} \Phi^{(1)}, \qquad x_3 = 0 \tag{89}$$

and the dynamic condition on the free surface results in the following (Kelvin) free surface condition:

$$(U^2 \partial_{x_1}^2 + g \partial_{x_3}) \Phi^{(1)} = 0, \qquad x_3 = 0. \tag{90}$$

On the wetted surface of the body the present derivation yields the impermeability condition:

$$N \cdot \partial_x \Phi^{(1)} = U N_1 - U \partial_{x_1} \eta^{(1)} \cdot N, \qquad x \in \partial D_{OW}. \tag{91}$$

The last term on the right-hand side of equation (91) is not found in the usual formulations of the Neumann-Kelvin problem. Its presence in equation (91) is a consequence of the method used here for the problem formulation. For slender ship hulls without excessive flare the contribution of this term should be negligible.

By changing assumption (87) to:

$$N_1 U = O(\Phi^{(0)}) \tag{92}$$

the formulation of the wave resistance problem is modified and instead of the Neumann-Kelvin formulation a Dawson type of the formulation is obtained. The pressure condition (77) on the free surface is changed to:

$$\eta_3^{(1)} = \frac{1}{g} [\frac{1}{2} (U^2 - | \partial_x \Phi^{(0)} |^2) - \partial_x \Phi^{(0)} \cdot \partial_x \Phi^{(1)}], \qquad x_3 = 0 \tag{93}$$

and the dynamic condition (79) on the free surface becomes:

$$\partial_x \Phi^{(0)} \cdot \partial_x (\partial_x \Phi^{(0)} \cdot \partial_x \Phi^{(1)}) - \partial_{x_3}^2 \Phi^{(0)} \partial_x \Phi^{(0)} \cdot \partial_x \Phi^{(1)} + g \partial_{x_3} \Phi^{(1)} =$$
$$\partial_x \Phi^{(0)} \cdot \partial_x [\frac{1}{2} (U^2 - | \partial_x \Phi^{(0)} |^2)] - \frac{1}{2} (U^2 - | \partial_x \Phi^{(0)} |^2) \partial_{x_3}^2 \Phi^{(0)}, \qquad x_3 = 0. \tag{94}$$

The impermeability condition on the reference wetted surface of the hull takes the form:

$$N \cdot \partial_x \Phi^{(1)} = (\partial_x \Phi^{(0)} \cdot \partial_x \eta^{(1)} - \eta^{(1)} \cdot \partial_x \partial_x \Phi^{(0)}) \cdot N, \quad x \in \partial D_{OW}. \tag{95}$$

The governing equations derived above for a Dawson type formulation of the wave resistance problem are close to other Dawson type formulations, such as for instance the governing equations presented in (Nakos and Sclavounos, 1991). The present formulation is different in two respects. It does not include certain terms dependent on $\partial_x \mid \partial_x \Phi^{(0)} \mid^2$ which are of $o(\Phi^{(1)})$ and therefore are negligible, owing to the assumption of small wave steepness. In addition, the expression on the right-hand side of the hull impermeability condition does not occur in the corresponding formula defined in (Nakos and Sclavounos, 1991). Similarily to the second non-homogeneous term in the hull impermeability condition (91), the non-homogeneous term in (95) results from the application of the explicit impermeability condition (78), and may be negligible for sufficiently slender hulls with nearly vertical sides.

Unless applied to very slender hulls, the Neumann-Kelvin and Dawson type formulations are in general not adequate to provide good quantitative predictions of ship wave resistance. A different family of formulations of the wave resistance problem are obtained by replacing the assumption (84) of small wave steepness with the assumption (Pawlowski, 1993):

$$\mid \partial_x \otimes \eta \mid = O(\Phi^{(0)}). \tag{96}$$

To distinguish it from condition (84) this condition is named the steep wave assumption. It is used here in conjunction with assumption (92). Equation (79) applied on the free surface, and the adopted assumptions lead to the following governing equations of the zero order:

$$g \partial_{x_3} \Phi^{(0)} + \frac{1}{2} \partial_x \Phi^{(0)} \cdot \partial_x (\mid \partial_x \Phi^{(0)} \mid^2) = 0, \quad x_3 = 0 \tag{97}$$

and the impermeability condition (86) on the reference wetted surface. It should be noticed that the zero order boundary value problem is non-linear. The first order governing equations are obtained as:

$$\begin{aligned} & g \partial_{x_3} \Phi^{(1)} + \frac{1}{2} \partial_x (\mid \partial_x \Phi^{(0)} \mid^2) \cdot \partial_x \Phi^{(1)} + \partial_x \Phi^{(0)} \cdot \partial_x (\partial_x \Phi^{(0)} \cdot \partial_x \Phi^{(1)}) \\ & + \partial_{x_3} [g \partial_{x_3} \Phi^{(0)} + \frac{1}{2} \partial_x \Phi^{(0)} \cdot \partial_x (\mid \partial_x \Phi^{(0)} \mid^2)] \eta_3^{(1)} = 0, \quad x_3 = 0 \end{aligned} \tag{98}$$

with:

$$\eta_3^{(1)} = \frac{-1}{g + \frac{1}{2} \partial_3 \mid \partial_x \Phi^{(0)} \mid^2} [\frac{1}{2} (\mid \partial_x \Phi^{(0)} \mid^2 - U^2) + \partial \Phi^{(0)} \cdot \partial_x \Phi^{(1)}], \quad x_3 = 0. \tag{99}$$

The first order impermeability condition on the reference wetted surface is given by equation (95). In this equation the formally zero order term which involves the wave slope is included because it is usually sufficiently small for slender and close to wall-sided hulls.

As shown by the examples of wave resistance computations provided below, the zero order steep wave formulation gives predictions of ship wave resistance close to the ones obtained with the use of Dawson type formulae. However, at least for hulls of moderate fullness the first order formulation shows results quantitatively close to experimental data.

## 3.3 Formulations for Unsteady Flow, Ship Motions in Waves

Unsteady flow problems of ship hydrodynamics arise from considering the interaction of a ship advancing with a constant mean forward speed in an ambient field of water waves. Interaction with the ambient waves causes the ship to perform oscillatory motions about its mean trajectory of forward motion. Mathematical modelling of the ship-wave interaction results in unsteady (transient) boundary value problems, unless regular ambient waves of small amplitude are assumed, and a linear, frequency domain solution is satisfactory. Here, an example is given of a non-linear, time domain formulation (Pawlowski, 1992), (Pawlowski and Bass, 1991). Using the method described above, a boundary value problem is formulated by considering the flow perturbation induced by the presence of a hull advancing in the flow of steep ambient waves.

For the ship moving with a mean forward speed U, a reference configuration of the hull is defined by its mean location relative to a correspondingly advancing coordinate system. This configuration of the hull and the undisturbed water surface at $x_3 = 0$ determine the reference fluid domain $D_0$. Assuming the hull to be sufficiently slender, the weak-scatterer hypothesis (Pawlowski, 1992) is imposed in the perturbation formulation by expressing the total velocity potential $\Phi$ in $D_0$ as a double perturbation series:

$$\Phi = -Ux_1 + \Phi^{(1,0)} + \Phi^{(0,1)} + \Phi^{(2,0)} + \cdots \tag{100}$$

where as before $x_1$ coordinate axis points in the direction of the mean forward velocity of the ship. Similarly, the domain displacement field is expressed by means of a series:

$$\eta = \eta^{(1,0)} + \eta^{(0,1)} + \eta^{(2,0)} + \cdots \tag{101}$$

The relative orders of magnitude in (100) are defined as:

$$\Phi^{(0,1)} = o(\Phi^{(1,0)}), \quad \Phi^{(2,0)} = o(\Phi^{(1,0)}) \tag{102}$$

where the symbol $o(.)$ denotes a quantity of a smaller order of magnitude than the quantity in brackets, and :

$$(\Phi^{(1,0)})^2 = O(\Phi^{(2,0)}), \quad \Phi^{(2,0)} = O(\Phi^{(0,1)}). \tag{103}$$

Analogous relations are applied to the terms in series (101). In the double series (100) the terms:

$$\Phi_W = \Phi^{(1,0)} + \Phi^{(2,0)} + .. \tag{104}$$

represent the velocity potential of the ambient wave field.

From relations (102) and (103) it is seen that the weak-scatterer hyptothesis consists in assuming that the disturbance induced in the ambient wave flow by the presence of the hull, which is represented by velocity potential $\Phi^{(0,1)}$, is significantly smaller than the disturbance to the calm water condition by the ambient wave flow represented by $\Phi_W$. However, the flow disturbance induced by the hull is not significantly smaller than the non-linear flow effects in the ambient wave field, which are represented by velocity potential $\Phi^{(2,0)}$.

Using equations (55) and (56) it is found that potential $\Phi_W$ satisfies the usual equations of the propagation of non-linear free surface waves. The disturbance potential $\Phi^{(0,1)}$ can be expressed as the sum of a steady, $\Phi_U$, and transient, $\Phi_S$, flow potentials:

$$\Phi^{(0,1)} = \Phi_U + \Phi_S \tag{105}$$

with:

$$\eta^{(0,1)} = \eta_U + \eta_S. \tag{106}$$

The steady potential satisfies the boundary conditions of the Neumann-Kelvin problem described above, if:

$$UN_1 = O(\Phi^{(0,1)}). \tag{107}$$

The unsteady potential is called the scattering potential, and satisfies the unsteady linear free surface condition:

$$[(\partial_t - U\partial_{x_1})^2 + g\partial_{x_3}]\,\Phi_S = 0, \quad x_3 = 0 \tag{108}$$

with the corresponding wave elevation determined by:

$$\eta_{S3} = -\frac{1}{g}(\partial_t - U\partial_{x_1})\Phi_S. \tag{109}$$

The impermeability condition on the reference wetted surface of the hull for the scattering potential is non-linear and has the form:

$$N \cdot \partial_x \Phi_S = v_n, \quad x \in \partial D_{0W} \tag{110}$$

with:

$$v_n = N \cdot (\partial_t - U\partial_{x_1})\eta_S + v_n' \tag{111}$$

and:

$$\begin{aligned} v_n' = [(\partial_t - U\partial_{x_1})(\eta^{(1,0)} + \eta^{(2,0)}) - (1 + \eta^{(1,0)} \cdot \partial_x)\partial_x \Phi^{(1,0)} \\ -(\partial_t - U\partial_{x_1})\eta^{(1,0)} \cdot \partial_x \eta^{(1,0)} + \partial_x \Phi^{(1,0)} \cdot \partial_x \eta^{(1,0)} - \partial_x \Phi^{(2,0)}] \cdot N. \end{aligned} \tag{112}$$

It follows that the weak-scatterer hypothesis is equivalent to the assumption:

$$v_n' = O(\Phi^{(0,1)}). \tag{113}$$

It is shown in (Pawlowski, 1992) that the right-hand side of equation (111) can be rewritten as:

$$v_n(x) = u_n(y) \doteq (v - u_W) \cdot n + U(n - N)_1 \tag{114}$$

including terms up to $O(\Phi^{(0,1)})$. In the above expression $v$ is the local velocity of the hull, relative to the reference configuration, $u_W$ is the absolute water velocity induced by the ambient wave field, and $n$ is the instantaneous normal vector on the hull surface.

The solution of the scattering boundary value problem is further facilitated, see (Pawlowski, 1992) and (Pawlowski and Bass, 1991), by expressing the normal velocity $u_n$ by means of a finite series of linearly independent and square integrable on the instantaneous wetted surface, normal shape functions $\psi_{Ni}(y)$:

$$u_n(y,t) = \sum_{i=1}^{m} \beta_i(t)\psi_{Ni}(y), \quad y \in \partial D_W \tag{115}$$

where $\beta_i(t)$ are functions of time. The normal shape functions are assumed to satisfy the Lipshitz condition on the hull surface, and to be invariant with respect to the hull motions. It is shown in (Pawlowski, 1992) that the impermeability condition (110) takes then the form:

$$N \cdot \partial_x \Phi_S = \sum_{i=1}^{m} \beta_i(t)\psi_{Ni}(x), \quad x \in \partial D_{0W}. \tag{116}$$

This last expression in conjunction with the free surface condition (108) allows one to solve the scattering problem, as a quasi-linear problem on the reference fluid domain. The solution can be obtained by the method of modal potentials, (Pawlowski, 1982), (Pawlowski and Bass, 1991), using appropriate Green's functions and initial conditions to solve the scattering problem. An example of the computation of wave induced loads on a ship hull, obtained with the use of the weak-scatterer hypothesis, is given below following (Pawlowski and Bass, 1991).

## 3.4 Formulations for Unsteady Flow, Evolution Equations and a Wave Radiation Condition

In an instantaneous fluid domain $D$ an unsteady flow with a free surface is governed by equations (48)-(53). The domain is divided by a control boundary $\partial D_C$ into a bounded control domain $D_C$ and a domain $D_e$ which extends horizontally to infinity. The control domain contains a floating body with a surrounding part of the free surface. In order to solve the flow problem at any point in time it is necessary to know the instantaneous location of the boundaries of the control domain and the instantanous values of either the normal velocity or velocity potential at all points on those boundaries. Equations which govern the change in time of the domain and its boundary data are called the evolution equations. By a sequential solution of the boundary value problem and integration of the evolution equations in time, starting from a given set of initial conditions, the solution of the unsteady flow problem is advanced in time. This solution is coupled with the solution of the equations of motion of the floating body which is subject to the generalized hydrodynamic forces exerted by the fluid on its instantaneous wetted surface.

The boundary data on the wetted surface of the body, $\partial D_W$, is determined from the motion of the body with the use of the impermeability condition (49). However, the determination of the instantaneous wetted surface must be performed taking into account both the motion of the body and of the free surface, $\partial D_F$. One possible approach is to trace the evolution of the free surface by following in time the motion of fluid particles on the free surface (Longuet-Higgins and Cokelet, 1976). Consequently the evolution equations obtained in this manner are called Lagrangian. Using condition (50) the Lagrangian evolution equations on the free surface take the form:

$$\frac{d}{dt}y = \partial_y \Psi, \quad y \in \partial D_F \tag{117}$$

and:

$$\frac{d}{dt}\Psi = -gy_3 + \frac{1}{2} \mid \partial_y \Psi \mid^2, \quad y \in \partial D_F. \tag{118}$$

Although the Lagrangian formulation imposes the least possible restrictions on the motion of the free surface, it also leaves the waterline on the body surface defined only implicitly as the intersection line of the instantaneous surface of the body $\partial D_B$ and free surface $\partial D_F$. This feature of the formulation and lack of control over the trajectories of the traced particles lead to difficulties in numerical applications.

An alternative formulation of the evolution equations on the free surface relies on a purely geometrical representation of $\partial D_F$, (Pawlowski, 1994). The locus of the free surface is defined as:

$$f(y,t) = y_3 - \zeta(y_1, y_2, t), \quad y \in \partial D_F \tag{119}$$

where $\zeta$ denotes the instantaneous free surface elevation above the $y_3 = 0$ datum. The locus equation gives the relation:

$$\partial_t f + \frac{d}{dt}y \cdot \partial_y f = 0, \quad y \in \partial D_F \tag{120}$$

from which, using (117), the following evolution equation of the free surface is obtained:

$$(\frac{d}{dt}y - \partial_y \Psi) \cdot (e_3 - \partial_y \zeta) = 0, \quad y \in \partial D_F. \tag{121}$$

This equation applies to the trajectory of a point located on any line which intersects the free surface. The trajectory of such a point can be defined by the equation:

$$\frac{d}{dt}y = \frac{d}{dt}\lambda \, \partial_\lambda(\eta_0 + \eta_B) + \partial_t \eta_B \tag{122}$$

where $\lambda$ is a parameter on the line, and the notation of equations (74) and (75) is used.

From equations (121) and (122) the following form of the evolution equation is derived:

$$\frac{d}{dt}\lambda = \frac{1}{\partial_\lambda(\eta_0 + \eta_B) \cdot (e_3 - \partial_y \zeta)}(\partial_y \Psi - \partial_t \eta_B) \cdot (e_3 - \partial_y \zeta), \quad y \in \partial D_F. \tag{123}$$

From equations (50) and (122) one obtains:

$$\frac{d}{dt}\Psi = -g\zeta - \frac{1}{2} \mid \partial_y \Psi \mid^2 + [\frac{d}{dt}\lambda \, \partial_\lambda(\eta_0 + \eta_B) + \partial_t \eta_B] \cdot \partial_y \Psi, \quad y \in \partial D_F. \tag{124}$$

The evolution equations (123) and (124) can be applied on the free surface and on the body surface, and therefore provide an explicit definition of the instantaneous waterline. The advantages of such an explicit formulation are confirmed by results of numerical simulations.

On the free surface one may choose fixed vertical lines with $\lambda = \zeta$ to define the trajectories of points on the free surface. The solution is then limited to a single valued wave elevation and excludes breaking waves. As a result equations (123) and (124) are simplified to:

$$\frac{d}{dt}\zeta = \partial_{y_3} \Psi - \partial_y \Psi \cdot \partial_y \zeta, \quad y \in \partial D_F \tag{125}$$

and:

$$\frac{d}{dt}\Psi = -g\zeta - \frac{1}{2} \mid \partial_y \Psi \mid^2 + \frac{d}{dt}\zeta \, \partial_{y_3} \Psi, \quad y \in \partial D_F. \tag{126}$$

Applications of equations (125) and (126) to two-dimensional flow problems are presented in (Pawlowski et al., 1991) and (Sen et al., 1990). For two-dimensional flow problems equations (125) and (126) were first used in (Faltinsen, 1978).

In order to complete the formulation of the unsteady boundary value problem it is necessary to define the evolution equations on the control boundary. Since the locus of the boundary is chosen as a part of the problem formulation, it is enough to determine the evolution of boundary data on $\partial D_C$. The boundary data are obtained by matching the flow in the control domain $\partial D_C$ with the flow in the external domain $D_e$ with the use of relations (52) and (53). In the application of those relations the potential of the external flow $\Psi_e$ is expressed as a transformed potential $\Phi_e$ which is defined on an external reference fluid domain $D_{e0}$. This external reference fluid domain is determined by the control boundary $\partial D_C$ and the undisturbed free surface $x_3 = 0$. The weak-scatterer hypothesis is adopted and the potential $\Phi_e$ and domain displacement field $\eta_e$ are expressed by means of double series analogous to those given by equations (100) and (101). This formulation includes the presence of an ambient wave field in the external domain, see equation (104).

The matching conditions are then obtained in the form:

$$\Psi = \exp(\eta \cdot \partial_x)(\Phi_{We} + \Phi_e^{(0,1)}), \quad y \in \partial D_C, \, x \in \partial D_{C0} \tag{127}$$

and:

$$N \cdot \partial_x \Phi_e^{(0,1)} = v_n, \quad x \in \partial D_{C0} \tag{128}$$

where $\partial D_{C0}$ denotes the reference control surface. In relation (128):

$$v_n(x) = u_n(y) \tag{129}$$

with:

$$u_n(y) = [\partial_y \Psi - \exp(\eta_e \cdot \partial_x)\, \partial \Phi_{We}] \cdot N, \quad y \in \partial D_C, \, x \in \partial D_{C0}. \tag{130}$$

The weak-scatterer hypothesis in this formulation consists in assuming the relation:

$$v_n = O(\Phi_e^{(0,1)}) \tag{131}$$

which is satisfied sufficiently far from the floating body for a three-dimensional problem. On the reference free surface the scattering potential $\Phi_e^{(0,1)}$ must fulfil the linear condition analogous to (108), with $U = 0$. The solution to this external scattering problem can be obtained in terms of the appropriate Green's function, (Pawlowski, 1992).

Similar to the governing equations, the evolution equations defined on the instantaneous control fluid domian, as described above, can be transformed to evolution equations on a corresponding reference control fluid domain by an application of the method of analytic mappings (Pawlowski, 1994). As a result non-linear and unsteady boundary value problems can be formulated and solved either directly on the instantaneous fluid domain (direct formulations and solutions) or on a reference fluid domain using perturbation series (perturbation formulations and solutions). As shown above by the example of the non-linear radiation condition, suitable combinations of the two approaches can also be applied. Consequently a significant flexibility and control over the used mathematical, numerical and algorithmic features of the solutions can be achieved.

# 4 Practical Numerical Modelling Techniques

## 4.1 Semi-discretization of The Boundary Value Problems by Means of Linear Boundary Elements

### 4.1.1 The Boundary Integral Formulation

Solutions to the potential flow problems discussed in the previous section are sought in terms of velocity potential $\Psi(y,t)$ which must satify the Laplace's equation on an instantaneous domain $D(t)$ together with appropriate boundary conditions on its boundary $\partial D(t)$. Equivalently it was shown that the problems can be reformulated in terms of a velocity potential $\Phi(x,t)$ defined on a reference domain $D_0(t)$ with corresponding boundary conditions on the reference boundary $\partial D_0(t)$. In this section numerical methods which are used by the authors to solve these problems in either formulation are discussed. For this purpose in a generic sense, $\Phi$, $D$ and $\partial D$ are used to denote either $\Psi(y,t)$ or $\Phi(x,t)$, $D(t)$ or $D_0(t)$, and $\partial D(t)$ or $\partial D_0(t)$, respectively. In general the boundary $\partial D$ consists of a free surface $\partial D_F$, an impermeable wetted surface $\partial D_W$, and a control surface $\partial D_C$ which represents a vertical cylindrical surface extending to the free surface $\partial D_F$. Below the free surface, $\partial D_C$ is closed by a horizontal surface element at a sufficient depth or an impermeable bottom. The free surface elavation is denoted by $\zeta$.

The application of Green's second identity to velocity potential $\Phi(P)$ and a fundamental solution $Y(P,Q)$ of the Laplace's equation, with point $P$ located in $D$ or on $\partial D$, results in the well-known formula for harmonic functions :

$$\Omega(P)\,\Phi(P) = \int_{\partial D} [Y(P,Q)\,\partial_{n_Q}\Phi(Q) - \Phi(Q)\,\partial_{n_Q}Y(P,Q)]\,dS(Q) \qquad (132)$$

with point $Q$ located on $\partial D$ and $n$ pointing outwards from $D$. Here $Y(P,Q)$ denotes the fundamental solution of Laplace's equation and is defined by:

$$Y(P,Q) = \ln\, r(P,Q) \qquad \text{for } 2-\text{D space}, \qquad (133)$$

and

$$Y(P,Q) = 1/r(P,Q) \qquad \text{for } 3-\text{D space}; \qquad (134)$$

$r(P,Q) = |x(P) - x(Q)|$ is the distance between points $P$ and $Q$; the subscript of $\partial_{n_Q}$ indicates the outward normal derivative taken at point $Q$. For $P$ located on $\partial D$, $\Omega(P)$ is the solid angle at point $P$ subtended inside $D$ by $\partial D - P$. This angle equals to $\pi$ ( in 2-D space) and $2\pi$ ( in 3-D space) when the normal $n$ to $\partial D$ is continuous at $P$. For $P$ located inside $D$, $\Omega(P)$ takes the value $2\pi$ and $4\pi$ for 2-D and 3-D spaces, respectively. The Cauchy Principal value of the integral over $\partial D$ is taken in (132).

The boundary integral formula expresses the potential $\Phi$ at any point $P$ inside $D$ in terms of its boundary values and those of its normal derivative. For point $P$ placed on $\partial_D$, and with Dirichlet, Neumann, or Robin (mixed) boundary conditions applied on non-overlapping parts of $\partial D$, the formula generates two coupled Fredholm integral equations, of the second kind for the unknown $\Phi$, and of the first kind with respect to the unknown $\partial_n\Phi$. It should be noted that the above integral equation (132) gives what is known as the direct formulation of the boundary value problems.

The integral equation (132) applies at any instant of time, and therefore for solutions advancing in time it is used in time stepping procedures for solving the transient problems. For unsteady problems, on the instantaneous wetted surface of the floating body the boundary data are determined by the impermeability condition . The unknown locus of the free surface and required boundary data on free surface $\partial D_F$ are obtained from initial conditions using the evolution equations. On the control boundary the boundary data is determined from a wave radiation condition. The evolution equations can be cast in a Lagrangian or Eulerian form, as described in Section (3.4).

### 4.1.2 Spatial discretization

Analytical solutions to the general non-linear, free surface potential flow problems are unavailable and numerical methods are used. It is assumed that $\partial D$ consists of $N_S$ piecewise smooth, open surfaces $\partial D_k$ such that:

$$\partial D = \bigcup_{k=1}^{N_S} \partial D_k. \tag{135}$$

The open surfaces $\partial D_k$ are further discretized into a finite number of non-overlapping triangular surface elements, each approximated by a plane triangle, $S_i^k$, such that:

$$\partial D_k = \bigcup_{i=1}^{M_k} S_i^k. \tag{136}$$

Collocation points $Q_i^k$ are chosen on each of the boundary elements $S_i^k$. The functions $\Phi$ and $\partial_n\Phi$ on $\partial D$ are approximated by a low order polynomial on each element in terms of the values determined at the corresponding collocation points $Q_i^k$. For a linear isoparametric representation of both the velocity potential and its normal derivative on the boundary surfaces, plane triangular elements are used together with collocation points placed at the vertices defining the surface mesh. The discretization scheme maintains a consistent order of approximation and represents an application of the first order boundary element method (Brebbia et al., 1984). For 2-D problems, the triangular elements are replaced by two-point straight line segments, and the collocation points coalesce at the end points of the elements.

For the steady boundary value problems described by equations (96) - (99), the free surface conditions are formulated using streamlines generated on the reference surface $x_3 = 0$ for the corresponding double-body problem (equations (82), (85) and (86)). The dividing streamline is formed by the coordinate axis $x_1$ and the intersection of the hull surface $\partial D_W$ with surface $x_3 = 0$. The set of streamlines on $x_3 = 0$ is generated from the solution to the equation:

$$\frac{dx_2}{dx_1} = \frac{u_2}{u_1} \tag{137}$$

where $dx_1$ and $dx_2$ are differential lengths on the plane $x_3 = 0$, $u_1$ and $u_2$ are the velocity components $\partial_{x_1}\Phi$ and $\partial_{x_2}\Phi$ of the double-body flow solution on $x_3 = 0$. The triangular discretization on $x_3 = 0$ is based on these streamlines and their density specified by $\Delta x_2$, and is a function of the forward speed under consideration. Contrary to reports in the literature, the procedure is found to be inexpensive, with no particular arrangements needed for the grid distribution, apart from an increase in the spacing between the streamlines the farther the distance from the dividing streamline (Baddour, et al., 1991).

The discretized form of the integral equation (132) can be written as:

$$\sum_{i=1}^{N_e} \int_{S_i} \Phi(Q)\, \partial_{n_Q} Y(P,Q)\, dS(Q) = \sum_{i=1}^{N_e} \int_{S_i} Y(P,Q)\, \partial_{n_Q}\Phi(Q)\, dS(Q), \tag{138}$$

where $S_i$ represents a boundary element $i$ and $N_e$ represents the total number of elements on the boundary mesh with $N_e = \sum_1^{N_S} M_k$. The integrals in equation (138) are over each boundary element and formally imply the inclusion of the special cases when $Q$ approaches point $P$. Since $\Phi(Q)$ and $\partial_n\Phi(Q)$ on element $i$ are linear functions of their values at the three vertices of the element, equation (138) can be re-written in terms of the values at the nodes in the form:

$$\sum_{j=1}^{N_n} \Phi(j) \int_{S(j)} \partial_{n(j)}[Y(P,Q)]\, T(j)\, dS(Q) = \sum_{j=1}^{N_n} \partial_{n(j)}\Phi(j) \int_{S(j)} Y(P,Q)\, T(j)\, dS(Q), \tag{139}$$

where $j$ represents the node number; $S(j)$ is the compact support of the tent function $T(j)$ that equals 1 at node $j$ and zero at other nodes of the meshed boundary; $\Phi(j)$ is the value of the velocity potential at node $j$; $n(j)$ the unit normal to the boundary at point $(j)$ directed outwards of the fluid; $N_n$ the total number of nodes on the meshed boundary $\partial D$.

The linear element approximation distributes the unknowns on the nodes of the boundary mesh and linearly vary the unknowns over each triangular element for 3-D problems (or line segment in 2-D formulations). The complexity of the linear element algorithm is higher than that of the constant element algorithms found in the literature as applied to the wave resistance problem. In these algorithms the collocation points take a central location on the boundary elements of the discretized boundary with the unknowns taking a constant value. In this case for 3-D problems quadrilateral panels are usually used. The disadvantages inherent in the use of constant element algorithms, such as the collocation points being located away from the actual boundary, the variables being discontinuous between panels, and the panelization of the boundary being inaccurate (except possibly for planar ones) are avoided in the present approach. Therefore, better computed results are expected when applying linear element approximations to 2 and 3-D problems.

Satisfying condition (139) at all the nodes of the boundary mesh, a linear system of equations can be obtained which is written in the form:

$$[A]\,\Phi = [B]\,\partial_n\Phi, \tag{140}$$

where $[A]$ and $[B]$ are $(N_n \times N_n)$ matrices of influence coefficients. The entries to these matrices are defined by:

$$a(i,j) = \int_{S(j)} \partial_{n(j)}[Y(P_i, Q)]\,T(j)\,dS(Q); \tag{141}$$

$$b(i,j) = \int_{S(j)} Y(P_i, Q)\,T(j)\,dS(Q) \tag{142}$$

where $P_i$ is a collocation point. When calculating the integrals in the formulae (141) and (142), careful consideration must be taken when the collocation points $i$ and $j$ coincide. Closed form formulae are derived for this case. The details are omitted from this presentation.

## 4.2 Computation of Transient Solutions

The spatial discretization scheme described above is applied to the boundary value problems described in Section (3) by identifying the open surfaces $\partial D_k$ with: the free surface $\partial D_F$, the bottom $\partial D_D$, the body wetted surface $\partial D_W$ and the control boundary $\partial D_C$. The simulation sought for is accomplished through a time marching procedure. Consistent initial conditions provide the boundary data on $\partial D$ at $t = 0$ : $\Phi$ on $\partial D_F$, $\partial_n \Phi$ on $\partial D_D$ and $\partial D_W$, and $\Phi$ on $\partial D_C$. From the solution of the appropriately rearranged system of linear equations, $\partial_n \Phi$ on $\partial D_F$ and $\partial D_C$, and $\Phi$ on $\partial D_D$ and $\partial D_W$ for that time instant are determined. The boundary conditions are then used by invoking appropriate evolution equations (as described in Section (3.4)) to obtain the required boundary data for the next time level. The evolution equations can be cast in a Lagrangian or Eulerian form. The Eulerian form may impose the condition of single valued wave elevation which must be satisfied at control boundary $\partial D_C$ to ensure an appropriate matching with the external flow. For example, the evolution of the wave elevation $\zeta$, and potential $\Phi$ on the instantaneous free surface can be followed at the image points on the undisturbed free surface, obtained by the projection of the collocation points along the vertical axis. The differential of the wave elevation at the image point:

$$d\zeta = \partial_t \zeta \, dt \tag{143}$$

and the variation of the potential at the points on the free surface, which undergo vertical displacements, is determined by:

$$d\Phi = \partial_t \Phi \, dt + \partial_{x_3} \Phi \, d\zeta \tag{144}$$

since $\Phi = \Phi(x_3, t)$ at these points, similarly for the data on the control boundary $\partial D_C$.

The evolution of the domain and the evolution of the boundary data are then expressed by two coupled equations of the form (see equations (125) and (126)):

$$\frac{d}{dt}\zeta = f(\partial_x \zeta, \partial_x \Phi, t) \tag{145}$$

$$\frac{d}{dt}\Phi = g(\zeta, \partial_x \zeta, \partial_x \Phi, t) \tag{146}$$

for known functions $f$ and $g$ and with $\frac{d}{dt}$ representing the time derivative, $\zeta$ the domain location and $\Phi$ the boundary data. These equations are integrated in time to determine the updated domain configuration $\partial D_{i+1}$ and the value of the updated boundary data $\Phi_{i+1}$ on the updated boundaries.

The present approach of following the evolution of the boundary $\partial D_F$ is different from the Lagrangian method utilized in a number of investigations of non-linear water waves, which are also based on boundary integral formulations. The Lagrangian method is capable of modelling multivalued free surface contours, making it an attractive approach to simulate extreme wave conditions as for example wave breaking. However, the Eulerian approach presented above provides several computational advantages. The image points on the free surface, which through the discretization are identified with collocation points, are not allowed to cluster, an adverse numerical effect inherent in the Lagrangian based methods. In addition, the collocation points cannot leave the computational domain at any time avoiding tracing such points. In particular the water line points on the wetted surface of the body are explicitly defined, as described in Section (3.4) and identified with collocation points. When applied in the most general form (equations (123) and (124)) the Eulerian formulation can also be used to model free surfaces of multivalued elevation.

On the body surface, $\partial_n \Phi$ on $\partial D_W$ is related to the body velocity by equation (65), which in turn is related to $\Phi$ through the body equations of motion and the Bernoulli's equation. The body equations of motion represent coupled ordinary differential equations in the body velocity and body displacement. In terms of the generalized velocity vector $v$ and displacement $x$ of the body they take the form:

$$\frac{d}{dt}v = h(F, J) \tag{147}$$

$$\frac{d}{dt}x = v \tag{148}$$

where $F$ represents the generalized resultant forces acting on the body and $J$ is the body tensor of inertia.

The evolution equations (145), (146), for the domain and its boundary data, and (147), (148) for the body motion can be put in the following general form:

$$\frac{dy}{dt} = G(y, t) \tag{149}$$

where $t$ denotes the time. A predictor-corrector algorithm based on a fourth order Adams-Bashforth-Moulton scheme is found to be convenient, resulting in convergent computations for the required integrations, with a uniform time stepping of size $\Delta t = t_{i+1} - t_i$ . The Adams-Bashforth algorithm has the explicit difference equation:

$$y_{i+1} = y_i + F_1(G_i, G_{i-1}, G_{i-2}, G_{i-3}, \Delta t) \tag{150}$$

for $i = 3, 4, \cdots$. The Adams-Moulton algorithm has the implicit difference equation:

$$y_{i+1}^{(k+1)} = y_i + F_2(G_{i+1}^{(k)}, G_i, G_{i-1}, G_{i-2}, \Delta t) \tag{151}$$

for $i = 2, 3, , \cdots$, and iterations $k = 0, 1, 2, \cdots$; with both equations having local truncation errors proportional to $y^{\mathrm{v}}(\Delta t)^4$ (Burden and Faires, 1985). The explicit formula is used in a predictor procedure, while the implicit one is used as a corrector. Improved approximations to $y_{i+1}^{(k+1)}$ are obtained by iterating the Adams-Moulton formula. In practice a finite number of iterations is required.

At each time step this scheme needs data at the preceding four steps. In the initiation of the solution the first three steps are treated by means of successively lower order schemes. It should be noted that other algorithms exist for the integration of equations of the form (149). Runge-Kutta schemes are popular for analogous initial value problems. For convenience as well as for consistency, the time integrations of the domain and boundary data, and of the of body motion use the same Adams-Bashforth-Moulton scheme. The calculation of $\partial_t \Phi$ pressure term in the hydrodynamic forces calculations is critical. The predicted values are computed by using backward difference scheme for $d\Phi/dt$ in the force evaluation, and an explicit scheme for integrating the equations of motion. At the corrector stage the scheme returns to the preceding time step correcting the force estimation by using a central difference scheme for $d\Phi/dt$, and the implicit scheme for the integration of the equations of motion. For the choice of the time step size, the Courant-Friedrichs-Lewy type of condition:

$$N_{CFL} = \left| c\frac{\Delta t}{\Delta x_1} \right| \leq 1 \tag{152}$$

with $c$ representing a wave celerity, is used. A stability analysis of a 2-D algorithm was carried out in (Pawlowski et al., 1991) and values of $N_{CFL}$ between 0.5 and 0.7 are found satisfactory.

# 5 Examples of Applications

Two examples of applications are presented here to illustrate the effectiveness of the theoretical and numerical tools described above. One is a computation of ship wave resistance, for a ship advancing with a constant forward speed in otherwise calm water. The other is a computation of hydrodynamic loads induced on a ship advancing in steep waves. Both problems are of fundamental importance in naval hydrodynamics. Solutions to the wave resistance problem are used in hull design and optimization from the point of view of ship propulsion and powering. Solutions to the problem of ship motions in waves are used for the prediction and evaluation of seakeeping qualities, motion stability and safety, and hull loads for design and operational purposes. Another feature shared by the two problems is that both of them are subjects of intensive research, in particular in the practically important range of hull shapes or operating conditions in which non-linear and/or transient flow phenomena are significant. In real terms both problems require better, theoretically more adequate, and numerically more robust solutions.

The derivations of novel formulations of the discussed boundary value problems, in the context and with the application of the analytic mapping method, were outlined in Section 3. In the wave resistance problem the novelty consists in adopting and implementing the assumption of steep waves, formula (96). In the ship motion problem the essence of the novelty is in the formulation and implementation of the weak-scatterer hypothesis, equation (113). The development and implementation of these formulations in a rational and consistent manner, relies on the use of the analytic mapping method. The numerical implementations depend also on the accuracy and robustness of the numerical tools.

The wave resistance problem is solved in three stages. A double-body flow problem is solved (see equations (82), (85) and (86)). This solution is used to create a streamline based grid on the free surface, for the remaining computations. Subsequently the non-linear zero order boundary value problem (equations (86) and (97)) is solved iteratively using the double-body solution as the first approximation. The solution of the zero order problem is then employed to solve the linear first order problem (equations (95), (98) and (99)). The computations were performed by a modified version of a simpler algorithm (Song et al., 1993), (Song and Baddour, 1993) for comparative purposes. The numerical model used the direct boundary element algorithm, with linear triangular elements, to solve the Laplace's equation, and a one-sided differencing algorithm along the streamlines of the double-body flow to implement the free surface boudary conditions (97) and (98). The impermeability condition (95) in these computations was used in the homogeneous form with the right-hand side set equal to zero. This simplification can be of relatively small significance for slender hulls.

The computations were carried out for a ship hull of moderate slenderness (Series 60, block coefficient 0.60) which is generally used as a refrence for the measurement and computation of ship wave resistance. Figures 1a, 1b and 1c, show some of the results of computations. Figure 1a displays the value of the non-dimensional wave resistance coefficient as obtained from computations by the discussed numerical model at zero and first order levels, and from computations by a numerical model based on a Dawson type algorithm. The results of computations are shown in comparison with experimental data gathered by a number of researchers at various research centres.

The comparison shows a good quantitative agreement of the first order computations with experimental data. The zero order computations are in good qualitative agreement with the experimental points and agree closely with the results obtained using the Dawson type algorithm. The convergence of the computed data to the experimental data with the increased order of approximation is consistent and correct over the whole range of computations. The strong convergence properties of the perturbation scheme are confirmed by the results of computations of the fluid velocities on the reference hull waterline. The horizontal and vertical components of the velocity are presented in Figures 1b and 1c respectively. It is seen that in both cases the first order components are an order of magnitude smaller than the zero order ones. In addition the computations demonstrated a perfect stability and convergence of the iterative solution of the zero order problem, and did not involve any numerical smoothing or damping techniques. For comparison with other methods of computation of wave resistance reference is made to (Kim and Lucas, 1990), (Raven, 1992), (Nakos and Sclavounos, 1991), and (Andrew et al., 1988).

As explained in Section 3, the theoretical model used in the ship motion example depends on an application of the weak-scatterer hypothesis and on the use of the modal potentials method (equations (115) and (116)). The details of the algorithm and extensive comparisons with experimental data of computed time histories of ship motions and fluid loads are presented in (Pawlowski and Bass, 1992). This reference also provides a discusion and comparison with other methods applicable to the computation of fluid loads and ship motions in waves.

Unlike for the numerical modelling of ship wave resistance, there are no model test results recognized as a reference for non-linear time domain computations of the loads and motions induced by steep waves on ships. A unique set of model test results (Grochowalski, 1989) is used in (Pawlowski and Bass, 1991). The model tests were performed with a model of a stern trawler in steep, breaking and close to breaking periodic waves. The model test results discussed here were obtained with a self propelled and manually controlled model of the trawler, advancing at a 30 deg angle relative to the direction of ambient wave propagation, with the waves slowly overtaking the model. The model performed large motions in the vertical and horizontal plane and was in danger of capsizing.

Figures 2a and 2b present computed time histories of the pitch moment (shown by the continuous lines) in comparison with the corresponding time history derived from the measured motions (shown by the dotted lines). The experimental data shows strong non-linear and transient features. It is compared in Figure 2a with a computed time history of the pitch moment which is induced by the undisturbed ambient wave flow (so called Froude-Krylov load). Figure 2b shows the same experimental time history compared with a computed time history of the pitch moment which includes the scattering effect. In both figures the computed time histories show a strong initial transient resulting from the lack of modelling of the unknown initial flow conditions. These parts of the computed records cannot be compared with the experimental data.

From the two figures it is seen that the very large differences between the Froude-Krylov and experimantal pitch moment are compensated by the addition of the scattering effect, so that the experimental and computed record in Figure 2b are in very good agreement, taking into account the nature of the comparison. This result confirms the applicability and effectivenes of the weak-scatterer hypothesis and of the computational algorithm based on it. Further results of computations and comparisons with the model test data can be found in (Pawlowski and Bass, 1991).

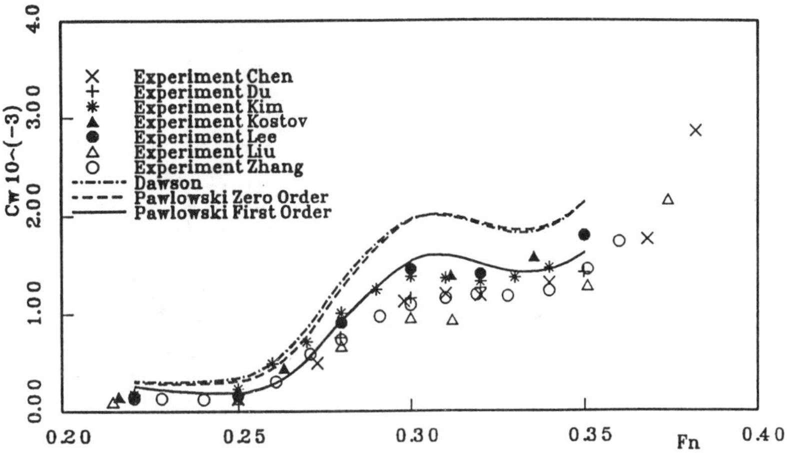


Fig. 1a Zero and First Order Non-linear Cw for Series 60 Ship Hull.

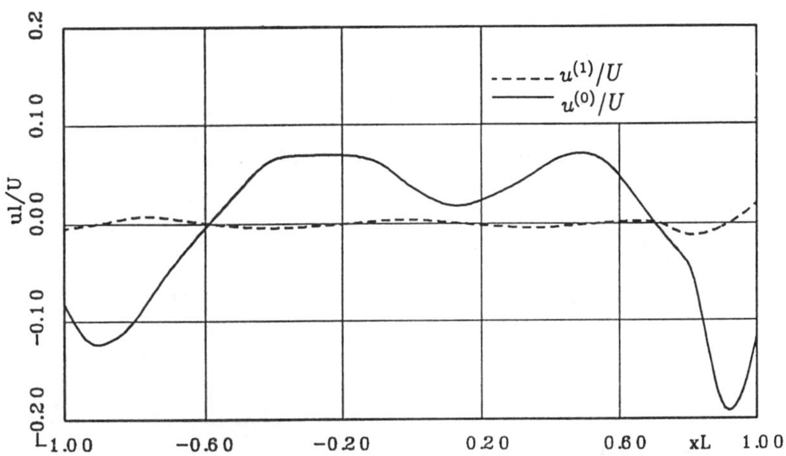


Fig. 1b Horizontal Fluid Velocity Along Series 60 Ship Hull (Fn = 0.25).

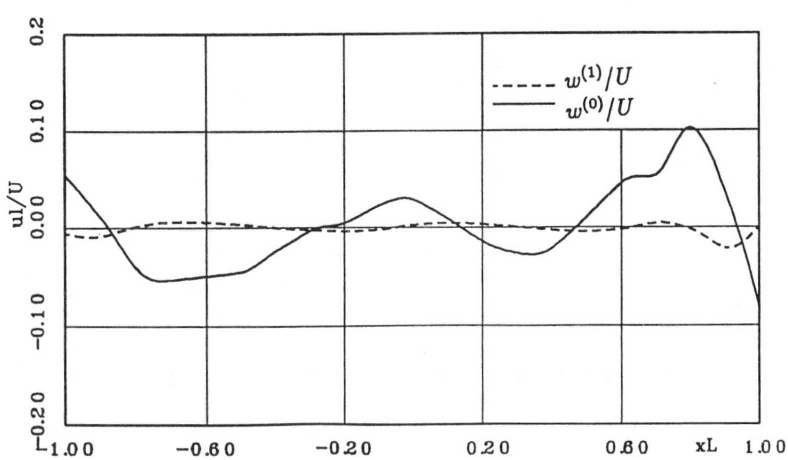


Fig. 1c Vertical Fluid Velocity Along Series 60 Ship Hull (Fn = 0.25).

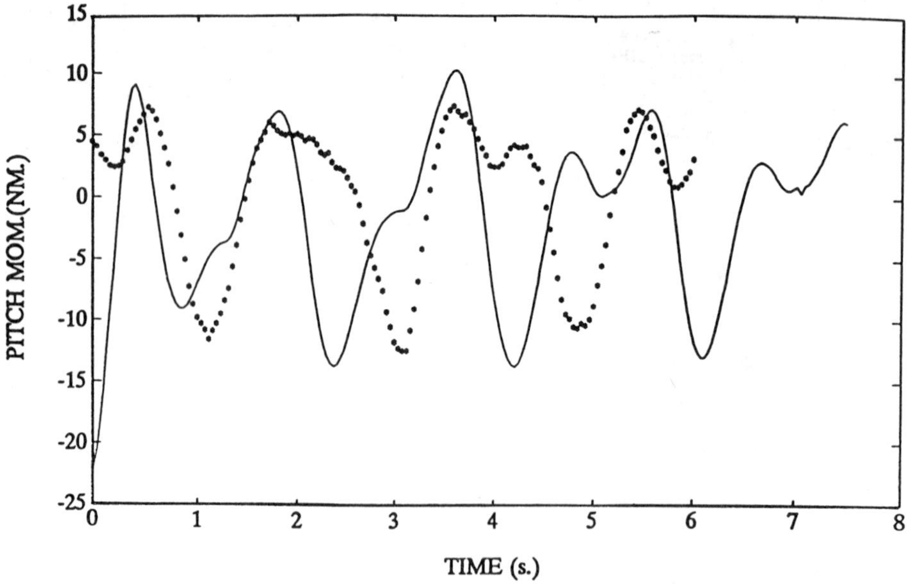


Fig. 2a TRAWLER, FREE RUNNING, FROUDE-KRYLOV LOADS.

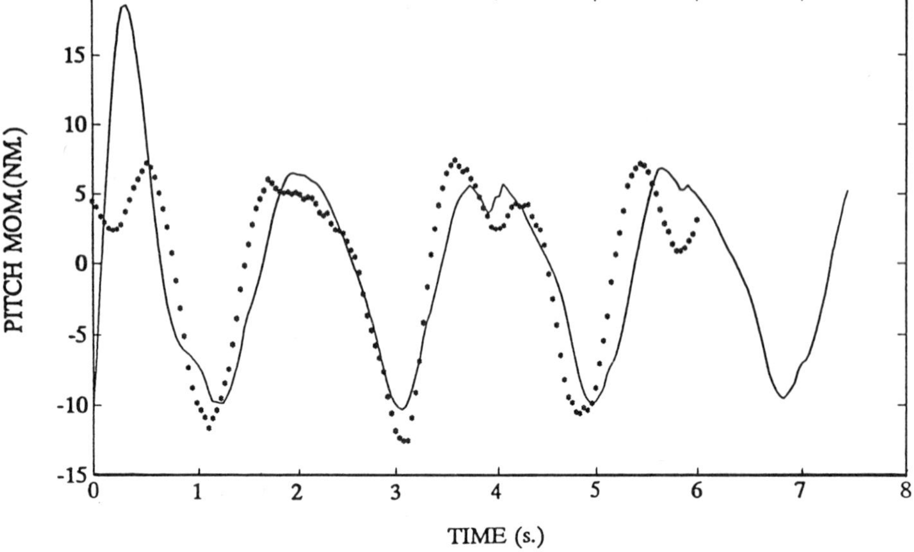


Fig. 2b TRAWLER, FREE RUNNING, HYDRODYNAMIC LOADS.

# 6 Conclusion

This chapter presents the modelling of potential flow problems of applied hydrodynamics with the use of the novel analytic mapping method. These problems involve the presence of an unknown free surface and impermeable boundaries, on which non-linear and time dependent boundary conditions are applied. The numerical techniques found to be suitable and reliable in obtaining numerical solutions to the steady and unsteady non-linear boundary value problems are described in this context, together with two specific applications. Novel problem formulations and computations of wave resistance of a ship hull advancing with a steady velocity in calm water, and of the non-linear time domain interaction of a ship with heavy seas are used as examples of applications. The developed methodology provides consistency and flexibility in the formulation and solution of unsteady, non-linear boundary value problems. It appears to open new possibilities for the numerical modelling of problems of marine hydrodynamics.

# References

Andrew, R.N., Baar, J.J.M. and Price, W.G. (1988) Prediction of Ship Wavemaking Resistance and Other Steady Flow Parameters Using Neumann-Kelvin Theory, *Trans. Roy. Inst. Naval Arch.*, Vol. 130, 119-129.

Baddour, R.E., Pawlowski, J.S. and Song, S.W. (1991) The Computation of Wave Patterns and Wave Resistance Induced by Submerged Bodies Advancing with Steady Forward Speed, *Symposium on Selected Topics of Marine Hydrodynamics*, NRC-C Marine Dynamics Conference, St. John's, NF,Canada,1-8.

Birkhoff, G. and Lynch, R.E. (1984) *Numerical Solution of Elliptic Problems*, SIAM, Philadelphia, p. 227.

Brebbia, C.A., Telles, J.C.F. & Wrobel, L.C. (1984) *Boundary Elements Techniques*, Springer-Verlag, New York.

Burden, R.L., Faires, J.D. (1985) *Numerical Analysis*, Third Edition, Prindle, Weber & Schmidt, Boston, Chap. 5.

Dieudonné, J. (1969) *Eléments d'Analyse, Tome I, Fondements de l'Analyse Moderne*, Gauthier-Villars, Paris, Chap. 9.

Faltinsen, O.M. (1978) A Numerical Non-linear Method of Sloshing in Tanks with Two-Dimensional Flow, *J. Ship Research*, Vol. 22, 3, 193-202.

Grochowalski, S. (1989) Investigation into the Physics of Ship Capsizing by Combining Captive and Free Running Model Tests, *Trans. SNAME*, Vol. 97, 169-212.

Kim, Y.H. and Lucas, T.R. (1990) Nonlinear Ship Waves, *Proc. 18th Symp. on Naval Hydrodynamics.*

Lebovitz, N.R. (1982) Perturbation Expansions on Perturbed Domains, *SIAM Review*, Vol. 24, 4, 381-400.

Longuet-Higgins, M.S. and Cokelet, E.D. (1976) The Deformation of Steep Surface Waves on Water: I. A Numerical Method of Computation,. *Proc. R. Soc. Lond., Series A*, Vol. 173, 683-707.

Nakos, D.E. and Sclavounos, P.D. (1991) Ship Motions by a Three-Dimensional Rankine Panel Methods, *Proc. 18th. Symp. on Naval Hydrodynamics*, 21-40.

Pawlowski, J.S. (1994) Elements of Non-linear Hydrodynamics, In preparation.

Pawlowski, J.S. (1993) A Perturbation Formulation of the Ship Wave Resistance Problem, *Eighth International Workshop on Water Waves and Floating Bodies*, NRC-C, St. John's, NF, Canada.

Pawlowski, J.S. (1992) A Non-linear Theory of Ship Motion in Waves, *The Nineteenth Symposium on Naval Hydrodynamics*, Seoul, Korea.

Pawlowski, J.S., Baddour, R.E. and Hookey, N. (1991) The Development of Second Generation Direct Solution Numerical Models of Large Motions of Floating Bodies in Waves, *Proc. Marine Dynamics Conference '91*, NRC-C, St. John's, NF, Canada, 43-49.

Pawlowski, J.S., and Bass, D.W. (1991) A Theoretical and Numerical Model of Ship Motions in Heavy Seas , *Trans. SNAME*, Vol. 99, 319-352.

Pawlowski, J.S. (1982) The Estimation of Diffraction Force Components from the Equivalent Motion Concept, *Int. Ship Building Progress*, Vol. 29, 62-73.

Raven, H.C. (1992) A Practical Non-linear Method for Calculating Ship Wavemaking and Wave Resistance, *Proc. 19th Symp. on Naval Hydrodynamics.*

Sen, D., Pawlowski, J.S., Lever, J. and Hinchey, M.J. (1990) Two-Dimensional Numerical Modelling of Large Motions of Floating Bodies in Waves, *Proc. Fifth Int. Conf. on Num. Ship Hydrodynamics*, Nat. Acad. Press, Washington, D.C., 351-373.

Song, S.W. and Baddour, R.E. (1993) Computations of Non-linear Ship Waves, *Eighth International Workshop on Water Waves and Floating Bodies*, NRC-C, St. John's, NF, Canada.

Song, S.W., Baddour, R.E. and Pawlowski, J.S. (1993) Nonlinear Computations of Wave Resistance for Surface Ships, *Proc. Second Canadian Marine Dynamics Conference*, UBC, Vancouver, BC, Canada, 1-8.

# Chapter 7

# Variational inequalities in physical oceanography

**M. Aslam Noor**

*Mathematics Department, College of Science, King Saud University, PO Box 2455, Riyadh 11451, Saudi Arabia*

## Abstract

In recent years, variational inequality theory has emerged as a fascinating and interesting branch of applicable mathematics, which enables us to consider a large number of unrelated linear and nonlinear problems arising in pure and applied sciences in a general and unified framework. In this paper, we show that the ideas and techniques of variational inequality theory can be extended to study a class of nonlinear obstacle problems in physical oceanography. These obstacle boundary value problems arise as a boundary layer problems from a model of large scale ocean circulation. Several computational techniques for computing the approximate solutions of the variational inequalities are given. We also suggest some open problems for further research with sufficient information and references.

## Keywords

Variational inequalities, auxiliary principle technique, Wiener-Hopf equations, projection method, iterative algorithms, convergence criteria, descent framework, odd order boundary problems.

## 1. Introduction

Oceanography is the study of the ocean making use of the various basic sciences including physics, chemistry, biology, geology and mathematics. Physical studies are carried out both by direct observations of the properties and movements and also by applying the basic physical principles of mechanics and thermodynamics to determine the motion. The observational approach is known as descriptive or synoptic oceanography. The second approach, called dynamical oceanography, is to use physical laws to endeavour to obtain mathematical relations between the forces acting on the ocean water and their consequent motions. Due to the complicated and nonlinear equations of state for sea water, many approaches including the boundary-layer technique have been developed in physical oceanography. The boundary-layer approach has been a very useful tool in analytic studies and has helped to improve our understanding of what dynamical effects may be important, even when the solutions are too idealized to be applied in detail to the ocean. In recent years Ierley and Ruehr [11] considered third order boundary values problems which arise as a model of large scale ocean circulation. Ierley and Ruehr [11] present extensive numerical calculations, whereas Dunbar [5] investigated the existence of solution of these third order boundary value problems using geometric techniques from qualitative differential equations. For the derivation of these problems, see Ierley and Ruehr [11] and the references therein.

This presentation involves the use of the variational principles and the variational inequalities, which helps us to study the odd-order obstacle boundary value problems. The variational principles can be used to interpret the basic principles of mathematical and physical sciences in the form of simplicity and elegance. These principles have played a significant and major role in the development of the general theory of relatively, gauge field theory in modern elementary particle physics and theory of solitone. The variational principles have been enriched by the discovery of the theory of variational inequalities, which is mainly due to Fichere [6] and Stampacchia [44]. The variational inequality theory is related to the simple fact that the minimum of the differentiable functional on a convex set in a normed space can be characterized by an inequality of the type $< F'(u), v - u > \geq 0$, where $F'(u)$ is the derivative of the functional $F[u]$. However, it is remarkable that the variational inequality theory allows many diversified applications. Variational inequality theory describes a broad spectrum of very interesting developments involving a link among various fields of mathematics, physics, geology, economics and engineering sciences.

Variational inequalities have been extended and generalized in many directions. A very useful and important generalization of variational inequalities is called the general variational inequality. This generalization is mainly due to Noor [23]. It has been shown [24,28,31] that the general variational inequality can be used to study both the odd - and even-order obstacle, free and moving boundary value problems. One of the most interesting and important problems in the variational inequality theory is the development of an efficient computable algorithm, see [2, 3, 4, 6, 8, 9, 10, 24, 33, 36]. Among the most effective numerical techniques are projection methods and its variant forms, linear approximation methods, penalty function method, and auxiliary principle techniques. The auxiliary principle technique has been used by Noor [32] to prove the existence of the solution of various types of variational inequalities. This approach deals with an auxiliary variational inequality problem and proving that the solution of the auxiliary problem is the solution of the original variational inequality problem. This technique is quite general and flexible. This technique can be used to formulate the variational inequality problem as a differentiable optimization problem.

This alternate formulation is very useful from a numerical point of view. In particular, we can develop a general framework for descent algorithms that solve the general variational inequalities. This technique includes many computational algorithms including the symmetrized Newton method, the Quasi-Newton methods, the linearized Jacobi method, the projection methods and its variant forms, decomposition and coordination.

In this paper, we show that the odd-order obstacle boundary value problems can be studied in the framework of the general variational inequalities. Our main goal is to present the material in the most transparent manner possible in order to help readers who work in other areas to learn as much as possible about the state-of-art in variational inequality theory. Thus we give the proofs in full. We hope that there will be enough details so that, say, an interested reader, who is motivated to read a little beyond this paper will be able to apply this material. In this paper, we shall give an account of certain developments in variational inequality theory which followed upon the discovery of general variational inequality in 1988 by Noor. The focus of our account will be recent glimmerings of understanding of the general variational inequalities in studying the odd order and nonsymmetric boundary value problems. We have tried to make the paper self-contained within its field but we consider it essential that readers interested in physical oceanography and variational inequalities should first aquaint themselves with the basic facts of Hilbert spaces and classical variational inequality theory.

## 2. Formulations And Basic Results

In this section, we prove that the minimum of the differentiable functionals associated with both the odd order and nonsymmetric boundary value problems can be characterized by a class of variational inequalities. We also show that a large number of odd order boundary value problems arising in physical oceanography can be studied in the general framework of these variational inequalities.

To be more precise, let $H$ be a real Hilbert space on which norm and inner product are denoted by $\| \cdot \|$ and $< \cdot, \cdot >$ respectively. Let $K$ be a nonempty closed convex set in $H$ and $f$ be a linear continuous functional on $H$. Given $T, g : H \longrightarrow H$ continuous operators, we consider the functional $I[v]$, defined as

$$I[v] = < Tv, g(v) > -2 < f, g(v) > \quad \text{for all } v \epsilon H. \tag{1}$$

The function $I[v]$, defined by (1) is called the general energy (potential, virtual work) functional. We remark that if $g(u) \equiv I$, the identity operator, then the functional $I[v]$ reduces to

$$J[v] = < Tv, v > -2 < f, v > \quad \text{for all } v \epsilon H, \tag{2}$$

which is known as the standard energy functional. It is well-known that if the operator $T : H \longrightarrow H$ is linear, symmetric and positive, then the minimum of the functional

$J[v]$, defined by (2), on the Hilbert space can be characterized by

$$< Tu, v >=< f, v > \quad \text{for all} \;\; v \epsilon H. \tag{3}$$

The formulation (3) is known as the weak formulation of the related boundary problems. This formulation has been used extensively not only to study the unique existence of the solution, but also to derive the finite element error estimates for the approximate solutions of the given boundary value problems, see [13] and the references cited therein.

It has been shown by Stampacchia [44] that if the operaktor $T$ is linear, symmetric and positive, then the minimum of $J[v]$, defined by (2), on the closed convex set $K$ in $H$ is equivalent to finding $u \epsilon K$ such that

$$< Tu, v - u >\geq< f, v - u > \quad \text{for all} \;\; v \epsilon K. \tag{4}$$

The inequality of the type (4) is known as the variational inequality, which was introduced by Stampacchia [44] and Fichera [6] in 1964 independently in different contexts. For the applications, numerical methods, sensitivity analysis and physical formulations of the variational inequalities, see [1, 3, 4, 9, 10, 12, 13, 14, 18, 20, 24, 25, 33, 36, 37, 40]. It is clear that the symmetry and positivity of the operator $T$ involved is must. On the other hand, there are many important problems, which are nonsymmetric and non-positive. For the nonsymmetric and odd order problems, many methods have been developed by several authors including Filippov [7], Petryshyn [37, 38] and Tonti [45] to construct the general energy functionals of the type (1) by introducing the concept of $g$-symmetry and $g$-positivity of the operator $T$. We need the following concepts.

**Definition 1.** For all $u, v \epsilon H$, an operator $T : H \longrightarrow H$ is said to be

(a). **g-symmetric** if

$$< Tu, g(v) >=< g(u), Tv >$$

(b). **g-positive** if

$$< Tu, g(u) >\geq 0$$

(c). **g-coercive (g-elliptic)** if there exists a constant $\eta > 0$ such that

$$< Tu, g(u) >\geq \eta \parallel g(u) \parallel^2$$

Note that $g$-coercivity implies $g$-positivity, but the converse is not true. It is also worth mentioning that there are operators, which are not $g$- symmetric, but $g$-positive. On the other hand, there are $g$-positive, but not $g$-symmetric operators. Furthermore, it is well-known [7] that if, for a linear operator $T$, there exists an inverse operator $T^{-1}$ on $R(T)$ with $\overline{R(T)} = H$, then one can find an infinite set of auxiliary operators $g$ such that the operator $T$ is both $g$-symmetric and $g$-positive.

We now consider the problem of finding the minimum of the functional $I[v]$, defined by (1), on the convex set $K$ in $H$, which is the main motivation of our next result.

**Theorem 1.** Let the operator $T : H \longrightarrow H$ be linear, $g$-symmetric and $g$-positive. If the operator $g : H \longrightarrow H$ is either linear or convex, then the function $u \epsilon H$ minimizes the functional $I[v]$ on the convex set $K$ in $H$ if and only if $u \epsilon H$ such that $g(u) \epsilon K$ and

$$< Tu, g(v) - g(u) > \geq < f, g(v) - g(u) >, \quad \text{for all } g(v) \epsilon K, \tag{5}$$

for all $v \epsilon H$.

**Proof.** Let $u \epsilon H$ such that $g(u) \epsilon K$ satisfy the inequality (5). Then using the $g$-positivity of the operator $T$, we have

$$\begin{aligned} < Tv, g(u) - g(v) > &\geq < f, g(v) - g(u) > + < Tv - Tu, g(v) - g(u) > \\ &\geq < f, g(v) - g(u) >, \quad \text{for all } g(v) \epsilon K. \end{aligned} \tag{6}$$

Since $K$ is a convex set, so for all $t \epsilon [0, 1], u, w \epsilon K$, $v_t = u + t(w - u) \epsilon K$. Taking $v = v_t$ in (6), and using the fact that $g$ is linear (convex), we have

$$< Tu, g(w) - g(u) > + t < t(w - u), g(w) - g(u) > \geq < f, g(w) - g(u) > . \tag{7}$$

We now define the function (see Noor [25]),

$$\begin{aligned} h(t) &= t < Tu, g(w) - g(u) > + \frac{t^2}{2} < T(w - u), g(w) - g(u) > \\ &\quad - < f, g(u) + t(g(w) - g(u)) >, \end{aligned}$$

such that

$$\begin{aligned} h'(t) &= < Tu, g(w) - g(u) > + t < T(w - u), g(w) - g(u) > - < f, g(w) - g(u) > \\ &\geq 0, \quad \text{by (7)}. \end{aligned}$$

Thus it follows that $h(t)$ is an increasing function on $[0, 1]$ and so $h(0) \leq h(1)$ gives us

$$< Tu, g(u) > -2 < f, g(u) > \leq < Tw, g(w) > -2 < f, g(w) >,$$

that is

$$I[u] \leq I[w],$$

which shows that $u \epsilon H$ minimizes the functional $I[v]$, defined by (1), on the convex set $K$ in $H$.

Conversely, assume that $u \epsilon H$ is the minimum of $I[v]$ on $K$, then

$$I[u] \leq I[v] \quad \text{for all } v \epsilon H \text{ such that } g(u) \epsilon K. \tag{8}$$

Taking $v = v_t \equiv u + t(w-u)\epsilon K$ for all $u, w\epsilon K$ and $t\epsilon[0,1]$ in (8), we have

$$I[u] \leq I[v_t].$$

Using (1) and the linearity (convexity) of $g$, we obtain

$$< Tu, g(w) - g(u) > + \frac{t}{2} < T(w-u), g(u) - g(w) > \geq < f, g(w) - g(u) >,$$

from which, as $t \to 0$, we have

$$< Tu, g(w) - g(u) > \geq < f, g(w) - g(u) >, \quad \text{for all } \; g(w)\epsilon K.$$

This completes the proof.

**Remark 1.** For the case $H = K$, Theorem 1 reduces to the well-known result of Petryshyn (38) which was proved under the assumption that $T$ is $g$-coercive. We remark that for $g \equiv I$, Theorem 1 reduces to the following well-known result in variational inequality theory, which is mainly due to Stampacchia [44].

**Theorem 2.** Let the operator $T : H \longrightarrow H$ be linear, symmetric and positive. The minimum of the functional $J[v]$, defined by (2), on the convex set $K$ in $H$ can be characterized by the variational inequality

$$< Tu, v - u > \geq < f, v - u > \quad \text{for all } \; v\epsilon K.$$

**Proof.** Its proof is similar to that of Theorem 1.

We now point out that a large number of both nonsymmetric and odd order obstacle, unilateral, free, moving and general equilibrium problems arising in various branches of pure and applied sciences can be formulated either in terms of (1) or in terms of (5). For illustration, we consider the obstacle boundary value problem of third order, which arises in physical oceanography.

Find $u$ such that

$$\left.\begin{array}{ll} \frac{-d^3u(x)}{dx^3} \geq f(x), & x\epsilon\Omega \\ u(x) \geq f(x), & x\epsilon\Omega \\ \left[\frac{-d^3u}{dx^3} - f(x)\right]\left[u(x) - \psi(x)\right] = 0, & x\epsilon\Omega \\ u(a) = 0, u'(b) = 0 = u'(a) & \end{array}\right\}, \tag{9}$$

where $\Omega = [a, b]$ is a domain, $\psi(x)$ and $f(x)$ are the given function. The function $\psi(x)$ is known as the obstacle function. The region, where $u(x) = \psi(x)$, for $x\epsilon\Omega$ is called the contact region (set).

We remark that the problem (9) is a generalization of the boundary value problem

$$\left.\begin{array}{r} \frac{-d^3u}{dx^3} = f(x), \quad x\epsilon\Omega \\ u(0) = 0, u'(a) = u'(b) = 0 \end{array}\right\}, \tag{10}$$

a problem considered by Dunbar [5]. The problems of the type (10) arise from a similarity solution of the so-called barotropic quasi-geostrophic potential vorticity equation for one layer ocean circulation. For the derivation of the equation and further references to the oceanography literature, see [5,11].

To study the problem (9) in the general framework of the variational inequality theory, we define

$$K = \left\{u\epsilon H_0^2(\Omega) : u(x) \geq \psi(x) \text{ on } \Omega\right\},$$

which is a closed convex set in $H_0^2(\Omega)$. For the definition and properties of the spaces $H_0^m(\Omega)$, see Kikuchi and Oden [13].

Using the technique of Tonti [45], we can easily show that the energy functional associated with the problem (9) is

$$\begin{aligned} I[v] &= \int_a^b \left(\frac{-d^3v}{dx^3}\right)\left(\frac{dv}{dx}\right)dx - 2\int_a^b f(x)\left(\frac{dv}{dx}\right)dx, \text{ for all } v\epsilon K \\ &= \int_a^b \left(\frac{d^2v}{dx^2}\right)\left(\frac{d^2v}{dx^2}\right)dx - 2\int_a^b f(x)\left(\frac{dv}{dx}\right)dx \\ &\equiv \; < Tv, g(v) > -2 < f, g(v) >, \end{aligned} \tag{11}$$

where

$$< Tu, g(v) >= \int_a^b \left(\frac{d^2u}{dx^2}\right)\left(\frac{d^2v}{dx^2}\right)dx \tag{12}$$

and

$$< f, g(v) >= \int_a^b f(x)\left(\frac{dv}{dx}\right)dx. \tag{13}$$

Here $g = \frac{d}{dx}$, the linear operator.

It is clear that the operator $T$ defined by the relation (12) is $g$- symmetric, $g$-positive and linear. Also we note that the operator $g = \frac{d}{dx}$ is a linear operator. Consequently all the assumptions of Theorem 1 are satisfied. Thus it follows from Theorem 1 that the minimum of the functional $I[v]$, defined by (11) is equivalent to finding $u\epsilon H$ such that $g(u)\epsilon K$ and the inequality (5) holds.

In fact, we conclude the problems equivalent to (9) are:

(a). **The Variational Problem**

Find $u \epsilon H_0^2(\Omega)$, which gives the minimum value to the functional

$$I[v] =< Tv, fg(v) > -2 < f, g(v) >$$

on the convex set $K$.

(b). **The Variational Inequality (Weak) Problem**

Find $u \epsilon H_0^2(\Omega)$ such that $g(u) \epsilon K$ and

$$< Tu, g(v) - g(u) > \geq < f, g(v) - g(u) > \quad \text{for all} \quad g(v) \epsilon K.$$

## 3. Existence Theory

In this section, we use the auxiliary principle technique to study the problem of the existence of a unique solution of a general variational inequality of which (5) is a special case. From now onwards, the operators $T, g : H \rightarrow H$ are nonlinear operators unless otherwise specified.

**Problem 1.** For given $T, g : H \longrightarrow H$ continuous operators, we consider the problem of finding $u \epsilon H$ such that $g(u) \epsilon K$ and

$$< Tu, g(v) - g(u) > \geq 0 \quad \text{for all} \quad g(v) \epsilon K, \tag{14}$$

which is known as the general nonlinear variational inequality. The problem (14) was introduced by Noor [23]. For $g \equiv I$, the identity operator, the problem (14) is equivalent to finding $u \epsilon K$ such that

$$< Tu, v - u > \geq 0, \quad \text{for all} \quad v \epsilon K. \tag{15}$$

The problem (15) was introduced and studied by Stampacchia [44] and Fichera [6] in 1964 independently. Since then, it has been shown that the ideas and techniques of variational inequality theory could be used to study many unrelated linear and nonlinear problems arising in pure and applied sciences in a general and unified framework.

We like to point out that if the set $K$ is not convex, then problem (14) could have no solution. Let $K = \{u : 2 \leq \| u \|_2 \leq 3\}$, which is closed but nonconvex set. Let $Tu = (u_2, -u_1)$, and $g \equiv I$, then the variational inequality $< Tu, v - u > \geq 0$ have no solution, where $H = R^n$. This shows that the convexity of the set cannot be relaxed.

We note that if $K^* = \{u \epsilon H; < u, v > \geq 0 \text{ for all } v \epsilon K\}$ is a polar cone of the convex cone $K$ in $H$ and $K \subset g(K)$, then it can be shown [12] that the problem (14) is equivalent to finding $u \epsilon H$ such that

$$g(u) \epsilon K, \; Tu \epsilon K^* \; \textit{and} \; < Tu, g(u) >= 0. \tag{16}$$

The problem (16) is known as the general nonlinear complementarity problem. For different choices of the operators $T, g$, and convex set $K$, one is able to obtain a number of known classes of the complementarity problems as special cases of the problem (16). For the recent applications, iterative methods, and physical formulation, see [10, 12, 20, 25, 33] and the references therein.

We need the following concepts and definitions.

**Definition.** An operator $T : H \longrightarrow$ is said to be

(i) **Strongly monotone** if there exists a constant $\alpha > 0$ such that

$$< Tu - Tv, u - v > \geq \alpha \parallel u - v \parallel^2 \quad \text{for all} \;\; u, v \epsilon H.$$

(ii) **Lipschitz continuous** if there exists a constant $\beta > 0$ such that

$$\parallel Tu - Tv \parallel \leq \beta \parallel u - v \parallel \quad \text{for all} \;\; u, v \epsilon H.$$

From (i) and (ii), it follows that $\alpha \leq \beta$.

**Theorem 3.** Let the operators $T, g : H \longrightarrow H$ be both strongly monotone and Lipschitz continuous. If $\gamma < \alpha$, where $\alpha$ is the strongly monotonicity constant and $\gamma$ is a constant defined by (22), then there exists a unique solution $u$ of the variational inequality (14).

**Proof. (a) Uniqueness**: Let $u_1 \neq u_2 \epsilon H$ be two solutions of the variational inequality (14). Then

$$< Tu_1, g(v) - g(u_1) > \geq 0 \quad \text{for all} \;\; g(v) \epsilon K \tag{17}$$

and

$$< Tu_2, g(v) - g(u_2) > \geq 0 \quad \text{for all} \;\; g(v) \epsilon K. \tag{18}$$

Taking $v = u_2$ in (17), $v = u_1$ in (18) and adding the resultant inequalities, we obtain

$$< Tu_1 - Tu_2, g(u_1) - g(u_2) > \leq 0,$$

which can be written as

$$< Tu_1 - Tu_2, u_1 - u_2 > \leq < Tu_1 - Tu_2, u_1 - u_2 - (g(u_1) - g(u_2)) > . \tag{19}$$

Since $T$ is strongly monotone, there exists a constant $\alpha > 0$ such that

$$\begin{aligned}
\alpha \parallel u_1 - u_2 \parallel^2 &\leq \; < Tu_1 - Tu_2, u_1 - u_2 > \\
&\leq \; < Tu_1 - Tu_2, u_1 - u_2 - (g(u_1) - g(u_2)) >, \quad \text{by (19)} \\
&\leq \; \parallel Tu_1 - Tu_2 \parallel\parallel u_1 - u_2 - (g(u_1) - g(u_2)) \parallel \\
&\leq \; \beta \parallel u_1 - u_2 \parallel\parallel u_1 - u_2 - (g(u_1) - g(u_2)) \parallel, \qquad (20)
\end{aligned}$$

where $\beta > 0$ is a Lipschitz continuity constant of $T$.

Now

$$\begin{aligned}
\parallel u_1 - u_2 - (g(u_1) - g(u_2)) \parallel^2 &\leq \; \parallel u_1 - u_2 \parallel^2 - 2 < g(u_1) - g(u_2), u_1 - u_2 > \\
&+ \; \parallel g(u_1) - g(u_2) \parallel^2 \\
&\leq \; (1 - 2\delta + \sigma^2) \parallel u_1 - u_2 \parallel^2, \qquad (21)
\end{aligned}$$

where $\delta > 0$ and $\sigma > 0$ are the strongly montotonicity and Lipschitz continuity constants of the operator $g$ respectively.

Combining (20) and (21), we obtain

$$\alpha \parallel u_1 - u_2 \parallel^2 \leq \beta \left(\sqrt{1 - 2\delta + \sigma^2}\right) \parallel u_1 - u_2 \parallel^2 = \gamma \parallel u_1 - u_2 \parallel^2,$$

where

$$\gamma = \beta \left(\sqrt{1 - 2\delta + \sigma^2}\right). \qquad (22)$$

Thus

$$(\alpha - \gamma) \parallel u_1 - u_2 \parallel^2 \leq 0,$$

from which it follows that $u_1 = u_2$, the uniqueness of the solution, since $\gamma < \alpha$.

(b) **Existence.** We use the auxiliary principle technique to prove the existence of a solution of (14). For a given $u \epsilon H$ such that $g(u) \epsilon K$, we consider the problem of finding a unique $w \epsilon H$ such that $g(w) \epsilon K$ satisfying the auxiliary variational inequality

$$< w, v - w > \geq < u, v - w > - \rho < Tu, g(v) - g(w) >, \text{ for all } g(v) \epsilon K, \qquad (23)$$

where $\rho > 0$ is a constant.

Let $w_1$ and $w_2$ be two solutions of (23) related to $u_1, u_2 \epsilon H$ respectively. It is enough to show that the mapping $u \longrightarrow w$ defined by (23) has a fixed point belonging to $H$ satisfying (23). In other words, it is sufficient to show that for well chosen $\rho > 0$

$$\| w_1 - w_2 \| \leq \theta \| u_1 - u_2 \|,$$

with $0 < \theta < 1$, where $\theta$ is independent of $u_1$ and $u_2$. Taking $v = w_2$ (respectively $w_1$) in (23) related to $u_1$(respectively $u_2$), we have

$$< w_1, w_2 - w_1 > \geq < u_1, w_2 - w_1 > -\rho < Tu_1, g(w_2) - g(w_1) >$$

and

$$< w_2, w_1 - w_2 > \geq < u_2, w_1 - w_2 > -\rho < Tu_2, g(w_1) - g(w_2) > .$$

Adding these inequalities, we have

$$\begin{aligned} < w_1 - w_2, w_1 - w_2 > &\leq < u_1 - u_2, w_1 - w_2 > -\rho < Tu_1 - Tu_2, g(w_1) - g(w_2) > \\ &= < u_1 - u_2 - \rho(Tu_1 - Tu_2), w_1 - w_2 > \\ &\quad -\rho < Tu_1 - Tu_2, w_1 - w_2 - (g(w_1) - g(w_2)) >, \end{aligned}$$

from which it follows that

$$\begin{aligned} \| w_1 - w_2 \|^2 &\leq \| u_1 - u_2 - \rho(Tu_1 - Tu_2) \| \| w_1 - w_2 \| \\ &\quad +\rho \| Tu_1 - Tu_2 \| \| w_1 - w_2 - (g(w_1) - g(w_2)) \| . \end{aligned} \tag{24}$$

Now using the strongly monotonicity and Lipschitz continuity of the operator $T$, we have

$$\begin{aligned} \| u_1 - u_2 - \rho(Tu_1 - Tu_2) \|^2 &\leq \| u_1 - u_2 \|^2 - 2\rho < Tu_1 - Tu_2, u_1 - u_2 > \\ &\quad +\rho^2 \| Tu_1 - Tu_2 \|^2 \\ &\leq (1 - 2\rho\alpha + \rho^2\beta^2) \| u_1 - u_2 \|^2 . \end{aligned} \tag{25}$$

Combining (21), (24), and (25), we have

$$\begin{aligned} \| w_1 - w_2 \| &\leq \left(\sqrt{1 - 2\rho\alpha + \rho^2\beta^2}\right) + \rho\left(\sqrt{1 - 2\delta + \sigma^2}\right) \| u_1 - u_2 \| . \\ &= \{\rho\gamma + t(\rho)\} \| u_1 - u_2 \| \\ &= \theta \| u_1 - u_2 \|, \end{aligned}$$

where $\theta = \rho\gamma + t(\rho)$, $\gamma = \beta\left(\sqrt{1-2\delta+\sigma^2}\right)$, and $t(\rho) = \sqrt{1-2\rho\alpha+\rho^2\beta^2}$.

We have to show that $\theta < 1$. It is clear that $t(\rho)$ assumes its minimum value for $\bar{\rho} = \frac{\alpha}{\beta^2}$ with $t(\bar{\rho}) = \sqrt{1-\frac{\alpha^2}{\beta^2}}$. For $\rho = \bar{\rho}, \rho k + t(\rho) < 1$ implies that $\rho k < 1$ and $0 < \rho < 2\frac{\alpha-\gamma}{\beta^2-\gamma^2}$. Thus it follows that $\theta < 1$ for all with

$$0 < \rho < 2\frac{\alpha-\gamma}{\beta^2-\gamma^2}, \quad \rho\gamma < 1 \text{ and } \gamma < \alpha,$$

showing that the map $u \longrightarrow w$ defined by (2.3) has a fixed point belonging to $H$, which is the solution of the variational inequality (14).

**Remark 2.** We note that if $g$ is linear or convex, then the auxiliary variational inequality problem (23) is equivalent to finding the minimum of the functional $I_1[w]$ on $K \subset H$, where

$$\begin{aligned} I_1[w] &= \frac{1}{2} < w, w > + \rho < Tu, g(w) > - < u, w >, \\ &= \frac{1}{2} < w-u, w-u > + \rho < Tu, g(w) - g(u) > \\ &\quad \rho < Tu, g(u) > - \frac{1}{2} < u, u >, \end{aligned}$$

which is the convex functional associated with the variational inequality (23). This equivalence has been used by Noor [24, 26, 32] to suggest a general algorithm for variational inequality (14) and related optimization problems. We note that if $w = u$, then $w$ is a solution of the variational inequality (14).

Following the ideas and techniques of Noor [29], we propose and analyze a general algorithm. For some $u \epsilon H$ such that $g(u) \epsilon K$, we introduce the following general auxiliary problem of finding the minimum of the functional $F[w]$, where

$$F[w] = \frac{1}{2} < A(w) - A(u), w - u > + \rho < Tu, g(w) - g(u) > .$$

Here $A : H \longrightarrow H$ is a continuous operator and $\rho > 0$ is a constant. If the operator $A$ is linear, symmetric, positive and the operator $T$ is linear, $g$-symmetric, and $g$-positive, then one can easily show that the minimum of the functional $F[w]$ on the convex set $K$ in $H$ can be characterized by the auxiliary variational inequality

$$< A(w), v - w > \geq < A(u), v - w > - \rho < Tu, g(v) - g(w) >, \text{ for all } g(v) \epsilon K.$$

It is obvious that if $w = u$, then $w$ is a solution of the variational inequality (14).

We note that in many applications the above auxiliary variational inequalities occur, which do not arise as a result of minimization problems. This motivates the interest of studying the above problems on its own, that is, without assuming a priori that these come out as an Euler inequality of an extremum problem. The main motivation of this section is to suggest a general auxiliary variational inequality problem which includes the above auxiliary problems as a special case.

For some $u \epsilon K$ such that $g(u) \epsilon K$, we consider the general auxiliary variational inequality problem of finding a unique $w \epsilon H$ such that $g(w) \epsilon K$ and

$$< A(w), v - w > \geq < A(u), v - w > -\rho < Tu, g(v) - g(w) >, \tag{26}$$

for all $g(v) \epsilon K$, where $\rho > 0$ is a constant and $A$ is a nonlinear operator on $H$.

It is clear that if $w = u$, then $w$ is a solution of the general variational inequality (14). Based on these observations, we now suggest and analyze the following iterative algorithm.

**Algorithm 1.**

(a). At $n = 0$, start with some initial $w_n$.

(b). At step $n$, solve the auxiliary problem (3.5) with $u = w_n$. Let $w_{n+1}$ denote the solution of the problem (14).

(c). If $\| w_{n+1} - w_n \| \leq \epsilon$, for given $\epsilon > 0$, stop. Otherwise repeat (b).

**Remark 3.** In this section, we have considered and studied a general auxiliary variational inequality problem, which includes the previously known ones as special cases. It is also shown that the auxiliary variational problem can be used to suggest an innovative and novel iterative algorithm for variational inequalities and related optimization problems. The auxiliary variational problem proposed in this section is quite general and flexible. The auxiliary variational principle and algorithm suggested may be extended for the multi valued and fuzzy operators involving variational inequalities.

## 4. Iterative Algorithms

In this section, we use the projection technique and its variant forms to prove that the variational inequality (14) is equivalent to the fixed point problems. This alternate formulation is very important from a numerical and approximation point of view and plays a fundamental part in suggesting a number of iterative algorithms for solving variational inequalities and complementarity problems. For this purpose, we need the following well-known [1] result.

**Lemma 1.** Let $K$ be a closed convex set in $H$. Then for a given $z \epsilon H$, $u \epsilon K$ satisfies

$$< u - z, v - u > \geq 0 \quad \text{for all } v \epsilon K,$$

if and only if $u = P_K z$, where $P_K$ is the projection of $H$ onto $K$. Furthermore, $P_K$ is a nonexpansive operator, that is,

$$\| P_K u - P_K v \| \leq \| u - v \| \quad \text{for all } \; u, v \epsilon H.$$

**Theorem 4.** Let $K$ be a closed convex set in $H$. Then $u \epsilon H$ such that $g(u) \epsilon K$ is a solution of the variational inequality (14) if and only $u \epsilon H$ satisfies the relation

$$g(u) = P_K \left[ g(u) - \rho T u \right],$$

where $\rho > 0$ is a constant.

**Proof.** Let $u \epsilon H$ such that $g(u) \epsilon K$ and (14) hold. Then

$$< Tu, g(v) - g(u) > \geq 0. \quad \text{for all } \; g(v) \epsilon K$$

can be written as

$$< g(u) - (g(u) - \rho T u), g(v) - g(u) > \geq 0.$$

where $\rho > 0$ is a constant.

Using Lemma 1, one can easily show that

$$g(u) = P_K \left[ g(u) - \rho T u \right].$$

the required result.

From Theorem 4, it is obvious that the variational inequality problem (14) is equivalent to finding the fixed point of the problem

$$u = F(u) \equiv u - g(u) + P_K \left[ g(u) - \rho T u \right],$$

which enables us to suggest the following iterative method.

**Algorithm 2.** For a given $u_0 \epsilon H$, compute $u_{n+1}$ by the iterative scheme

$$u_{n+1} = u_n - g(u_n) + P_K \left[ g(u_n) - \rho T u_n \right], \quad n = 0, 1, 2, ... \tag{27}$$

$\rho > 0$ is a constant.

For the convergence analysis of Algorithm 2, see Noor [23].

Let $P_K$ be the projection of $H$ into $K$ and $Q_K = I - P_K$, where $I$ is the identity operator. For given operators, $T, g : H \longrightarrow H$, we consider the problem of finding $z \epsilon H$ such that

$$T g^{-1} P_K z + \rho^{-1} Q_K z = 0, \tag{28}$$

where $\rho > 0$ is a constant and $g^{-1}$ is the inverse of the operator $g$. The equations of the type (29) are called the general Wiener- Hopf equation. For $g \equiv I$, the Wiener-Hopf equation (28) reduce to

$$TP_K z + \rho^{-1} Q_K z = 0, \tag{29}$$

which is known as the Wiener-Hopf equation. For the general treatment and applications of the Wiener-Hopf equations, see Speck [43] and the references therein. Shi [41] has proved that the variational inequality (15) is equivalent to solving the Wiener-Hopf equation (29) using Lemma 1. We extend the technique of Shi [41] to prove that the general variational inequality (14) is equivalent to the general Wiener-Hopf equations (28).

**Theorem 5**. The general variational inequality (14) has a solution $u \epsilon H$ such that $g(u) \epsilon K$ if and only if the general Wiener-Hopf equation (28) has a solution $z \epsilon H$, where

$$g(u) = P_K z, \tag{30}$$

and

$$z = g(u) - \rho T u, \tag{31}$$

where $\rho > 0$ is a constant and $g^{-1}$ exists.

**Proof.** Let $u \epsilon H$ such that $g(u) \epsilon K$ be a solution of the variational inequality (14). Then by theorem 4, it follows that

$$g(u) = P_K \left[ g(u) - \rho T u \right].$$

Using the fact $Q_K \equiv I - P_K$ and the above inequality repeatedly, we have

$$\begin{aligned} Q_K \left[ g(u) - \rho T u \right] &= g(u) - \rho T u - P_K \left[ g(u) - \rho T u \right] \\ &= -\rho T u \\ &= -\rho T g^{-1} P_K \left[ g(u) - \rho T u \right], \end{aligned} \tag{32}$$

from which it follows that

$$T g^{-1} P_K z + \rho^{-1} Q_K z = 0,$$

where

$$z = g(u) - \rho T u$$

Conversely, let $z \epsilon H$ be a solution of (28), then we have

$$\rho T g^{-1} P_K z = -Q_K z = P_K z - z. \tag{33}$$

Now from (33) and Lemma 1, for all $g(v) \epsilon K$, we obtain

$$0 \leq < P_K z - z, g(v) - P_K z > = < \rho T g^{-1} P_K z, g(v) - P_K z >,$$

from which, it follows that

$$< Tg^{-1}P_K z, g(v) - P_K z > \geq 0, \quad \text{for all } g(v) \epsilon K.$$

Thus $g(u) = P_K z$ is a solution of (14) and from (33), we have

$$z = g(u) - \rho T u,$$

the required result.

Theorem 5 implies that the general variational inequality problem (14) is equivalent to solving the general Wiener-Hopf equation (28). This equivalence has been used to suggest and analyze a number of iterative methods for solving variational inequalities and complementarity problems.

**I.** The Wiener-Hopf equation (28) can be written as

$$Q_K z = -\rho T g^{-1} P_K z,$$

which implies that

$$\begin{aligned} z &= P_K z - \rho T g^{-1} P_K z \\ &= g(u) - \rho T u, \quad \text{using (31).} \end{aligned}$$

This formulation enables us to suggest the following iterative schemes.

**Algorithm 3.** For a given $z_0 \epsilon H$, compute $z_{n+1}$ by the iterative schemes

$$g(u_n) = P_K z_n$$

and

$$z_{z+1} = g(u_n) - \rho T u_n, \quad n = 0, 1, 2, \ldots$$

**II.** By an appropriate rearrangement, the Wiener-Hopf equation (28) can be written in the following form

$$\begin{aligned} z &= P_K z - \rho T g^{-1} P_K z + (I - \rho^{-1}) Q_K z \\ &= g(u) - Tu + (I - \rho^{-1}) Q_K z, \quad \text{using (31),} \end{aligned}$$

which is another equivalent fixed point problem. Using this formulation, we suggest the following iterative method.

**Algorithm 4.** For a given $z_0 \epsilon H$, compute $z_{z+1}$ by the iterative scheme

$$g(u_n) = P_K z_n$$

and

$$z_{n+1} = g(u_n) - Tu_n + (I - \rho^{-1})Q_K z_n, \quad n = 0, 1, 2, ...$$

**III.** If the operator $T$ is linear and $T^{-1}$ exists, then the Wiener-Hopf equation (28) can be written as

$$z = \left(I - \rho^{-1} g T^{-1}\right) Q_K z.$$

This formulation enables us to suggest the following iterative scheme.

**Algorithm 5.** For a given $z_0 \epsilon H$, compute $z_{z+1}$ by the iterative scheme

$$z_{z+1} = \left(I - \rho g T^{-1}\right) Q_K z_n, \quad n = 0, 1, 2, ...$$

**Remark 4.** We like to remark that if $g \equiv I$, the identity operator, then we obtain various algorithms for solving variational inequalities of the type (15), which are mainly due to Noor [27] and Shi [41]. For the convergence analysis of these Algorithms see Noor [31]. We like to point out that for $g \equiv I$, the identity operator, Algorithm 5 has been used to compute the numerical solutions of the obstacle and unilateral problems by Pitonyak, Shi and Shillor [39]. The results so far obtained are very encouraging. There is still no comparison among various iterative methods for solving the obstacle and unilateral problems. This is an open problem and offers an interesting and facinating area for further research.

**Remark 5.** If the operator $T$ is monotone and $g \equiv I$, the identity operator, then the projection Algorithm 2 need not converge. For example, let $K = \{u : \| u \|_2 \leq 1\}$, $Tu = (u_2 - u_1)$ and $H = R^n$. Clearly the set $K$ is convex and the variational inequality $< Tu, v - u > \geq 0$ has the unique solution $u = 0$. On the otherhand, if the projection method is initialized on the boundary, it always stays on the boundary and does not converge. Also if it is initialized at any non-zero interior point, it moves toward the boundary. In this case, using the technique of Korpelevich [15], and Luo and Tseng [18], we suggest the following iterative methods.

**Algorithm 6.** Given $u_0 \epsilon H$ such that $g(u_0) \epsilon K$, compute $u_{n+1}$ by the iterative scheme

$$u_{n+1} = \tilde{u}_n - g(\tilde{u}_n) + P_K\left[g(u_n) - \rho T\, \tilde{u}_n\right]$$

and

$$\tilde{u}_n = u_n - g(u_n) + P_K\left[g(u_n) - \rho T u_n\right], \quad n = 0, 1, 2, ...$$

where $\rho > 0$ is a constant.

**Algorithm 7.** For given $u_0 \epsilon H$ such that $g(u_0) \epsilon K$, compute $u_{n+1}$ by the iterative scheme

$$u_{n+1} = u_n - g(u_n) + P_K[u_n - \rho T u + e],$$

where $\rho > 0$ is a constant and $e$ is a sufficiently small "error" vector depending on $u$. This is a broad class that includes Algorithm 2 and Algorithm 6 as special cases. The quantities $\rho$ and $e$ may depend upon $u$ and may be viewed as algorithm parameters whereby different choices of $\rho$ and $e$ lead to different algorithms.

In many cases, we cannot compute $Tu$, but we may have access to $Tu + w$, $w$ is a random variable representing measurement or observation noise. One could still try to use Algorithm 2, which gives rise to the following method.

**Algorithm 8.** For a given $u_0 \epsilon H$ such that $g(u_0) \epsilon K$, compute $u_{n+1}$ by the iterative scheme

$$u_{n+1} = u_n - g(u_n) + P_K\left[g(u_n) - \rho(Tu_n - w_n)\right], \quad n = 0, 1, 2, ...$$

where $w_n$ is the noise in the measurement of $Tu$.

For a given monotone operator $T$, we consider the regularization $T_\epsilon$ of $T$ as

$$T_\epsilon = T + \epsilon I, \tag{34}$$

$I$ is the identity operator and $\epsilon > 0$ is a parameter.

In this case, the variational inequality problem (14) is equivalent to finding $u \epsilon H$ such that $g(u) \epsilon K$ and

$$< T_\epsilon u, g(v) - g(u) > \geq 0, \quad \text{for all } g(v) \epsilon K. \tag{35}$$

The variational inequality (35) is known as the perturbed variational inequality. It is clear that the operator $T_\epsilon$ defined by (34) is strongly monotone and Lipschitz continuous, so the perturbed variational inequality (35) has a unique solution see Theorem 3. From Lemma 1, it follows that the solution $u \epsilon H$ such that $g(u) \epsilon K$ satisfying (35) can be characterized by the fixed point problem

$$\begin{aligned} u &= F(u) \equiv u - g(u) + P_K\left[g(u) - \rho T_\epsilon u\right] \\ &= u - g(u) + P_K\left[g(u) - \rho T u - \rho \epsilon u\right], \text{ by (34).} \\ &= u - g(u) + P_K\left[(1 - \rho\epsilon)u - \rho T u + g(u) - u\right]. \end{aligned} \tag{36}$$

This fixed point formulation enables us to suggest and analyze the following algorithm.

**Algorithm 9.** For a given $u_0 \epsilon H$ such that $g(u_0) \epsilon K$, compute

$$u_{n+1} = u_n - g(u_n) + P_K\left[(1 - \rho\epsilon)u_n - \rho T u_n + g(u_n) - u_n\right], \; n = 0, 1, 2, ... \tag{37}$$

where $\rho > 0$ is a constant and $\epsilon > 0$.

We now study those conditions under which the approximate solution $u_{n+1}$ obtained from (37) converges to the exact solution $u$ of the variational inequality (35).

**Theorem 6.** Let $T : H \longrightarrow H$ be a monotone Lipschitz continuous operator. If $g : H \longrightarrow H$ is a strongly montone Lipschitz continuous operator, then

$$u_{n+1} \longrightarrow u \quad \text{strongly in } H,$$

for

$$\left|\rho - \frac{\epsilon}{\epsilon^2+\beta^2}\right| < \frac{\sqrt{\epsilon^2-(2k-k^2)(\epsilon^2+\beta^2)}}{\epsilon^2+\beta^2}, \ \epsilon > \frac{\beta\sqrt{2k-k^2}}{1-k} \text{ and } k<1,$$

where $u_{n+1}$ satisfies the relation (37) and $u$ is the solution of (35).

**Proof.** Let $u\epsilon H$ such that $g(u)\epsilon K$ satisfy the relation (35). Since the solution of (35) can be characterized by the relation (36). Hence from (36) and (37), we obtain

$$\begin{aligned} \| u_{n+1} - u \| &\le \| u_n - u - (g(u_n) - g(u)) \| + \| P_K\left[(1-\rho\epsilon)u_n - \rho T u_n + g(u_n) - u_n\right] \\ &\quad - P_K\left[(1-\rho\epsilon)u - \rho T u + g(u) - u\right] \| \\ &\le 2 \| u_n - u - (g(u_n) - g(u)) \| \\ &\quad + \| (1-\rho\epsilon)(u_n - u) - \rho(Tu_n - Tu) \| . \end{aligned} \tag{38}$$

Since $T$ is a monotone Lipschitz continuous operator, so

$$\begin{aligned} \| (1-\rho\epsilon)(u_n-u) - \rho(Tu_n - Tu) \|^2 &\le (1-\rho\epsilon^2) \| u_n - u \|^2 + \rho^2 \| Tu_n - Tu \|^2 \\ &\quad -2\rho(1-\rho\epsilon) < Tu_n - Tu, u_n - u > \\ &\le ((1-\rho\epsilon)^2 + \rho^2) \| u_n - u \|^2 . \end{aligned} \tag{39}$$

Similarly, using the strongly monotonicity and Lipschitz continuity of the operator $g$, one can have

$$\begin{aligned} \| u_n - u - (g(u_n) - g(u)) \|^2 &\le \| u_n - u \|^2 - 2 < g(u_n) - g(u), u_n - u > \\ &\quad + \| g(u_n) - g(u) \|^2 \\ &\le (1 - 2\sigma + \delta^2) \| u_n - u \| . \end{aligned} \tag{40}$$

Combining (38), (39) and (40), we obtain

$$\| u_{n+1} - u \| \leq \left\{ 2\left(\sqrt{1-2\sigma+\delta^2}\right) + \sqrt{(1-\rho\epsilon)^2+\rho^2\beta^2} \right\} \| u_n - u \|$$

$$= \{k + t(\rho)\} \| u_n - u \|$$

$$= \theta \| u_n - u \|,$$

where

$$\theta = k + t(\rho),\ k = 2\sqrt{1-2\sigma+\delta^2} \text{ and } t(\rho) = \sqrt{(1-\rho\epsilon)^2+\rho^2\beta^2}.$$

We now have to show that $\theta < 1$. It is clear that $t(\rho)$ assumes its minimum value for $\bar{\rho} = \frac{\epsilon}{\epsilon^2+\beta^2}$. For $\rho = \bar{\rho}$, $k + t(\rho) < 1$ implies that $k < 1$. Thus it follows that $\theta < 1$ for all $\rho$ with

$$\left| \rho - \frac{\epsilon}{\epsilon^2+\beta^2} \right| < \frac{\sqrt{\epsilon^2 - (2k-k^2)(\epsilon^2+\beta^2)}}{\epsilon^2+\beta^2},\ \epsilon > \frac{\beta\sqrt{2k-k^2}}{1-k} \text{ and } k < 1.$$

Since $\theta < 1$, it follows that the mapping defined by (36) has a fixed point such that $u \epsilon H$, such that $g(u) \epsilon K$ satisfying (35). Furthermore, it also implies that $u_{n+1} \longrightarrow u$ strongly in $H$. This completes the proof.

## 5. Equivalence

In the previous sections, we have discussed many numerical methods for solving the variational inequalities (14) and related optimization problems. In fact, we have studied quite different techniques for variational inequalities,namely the auxiliary principle technique and the projection methods. Here we prove that the auxiliary principle technique includes the projection methods and its variant forms as special cases. To be more precise, let $g \equiv I$, the identity operator, then the auxiliary functional $I_1[w]$ reduces to

$$I_2[w] = \frac{1}{2} < w - u, w - u > + \rho < Tu, w - u >, \tag{41}$$

which is a quadratic differential convex functional. It can be shown that the minimum of $I_2[w]$ on $K \subset H$ is equivalent to finding a unique $w \epsilon K$ such that

$$< w, v - w > \geq < u, v - w > - \rho < Tu, v - w >, \tag{42}$$

which is equivalent to finding the fixed point problem

$$w = P_K[u - \rho Tu], \tag{43}$$

by using Lemma 1. It can be easily shown that the fixed point problem has a unique fixed point $w = u\epsilon K$ satisfying the variational inequality (15). Also from Theorem 3 and Theorem 5, we conclude the variational inequalities (15) are equivalent to the fixed point problem and the Wiener-Hopf equation (28).

From the above observations and discussions, it is clear that the auxiliary principle technique includes the projection methods and its variant forms (the Wiener-Hopf equations) as special cases. We would like to remark that the projection method and its variant forms cannot be applied to develop iterative methods for finding the approximate solutions of the variational inequalities of the mixed type as shown by Noor [29, 32]. Secondly, it is very difficult to find the projection of the whole space on the convex set except in simple cases. However, the auxiliary variational principle can be effectively used to study the problem of the existence of solution of mixed type variational inequalities. It has been shown by Noor [29, 32], that the auxiliary principle technique is equivalent to finding the minimum of the associated auxiliary functional on the convex set in the space.

In 1992, Fukushima [8] proved the general asymmetric variational inequality can be formulated as a differentiable optimization problem using the projection method. Namely we define the mapping $M(u) = P_K[u - \rho Tu]$. Then it is known [1, 14] that $u\epsilon K$ is a solution of (15) if and only if $u$ is the fixed point of the mapping $M$, that is, $u = M(u)$. Using this alternate formulation Fukushima [8] proved that the variational inequality problem (15) is equivalent to the optimization problem of finding the minimum of $f(x)$ over $K \subset H$, where

$$f(u) = + < Tu, M(u) - u > + \frac{1}{2} < M(u) - u, M(u) - u > . \tag{44}$$

We now prove that the functional $f(u)$ is exactly the auxiliary functional $I_2[w]$, where

$$\begin{aligned} I_2[w] &= \frac{1}{2} < w, w > - < u, w > + \rho < Tu, w > \\ &= \frac{1}{2} < w - u, w > - \frac{1}{2} < u, w > + \rho < Tu, w > \\ &= \frac{1}{2} < w - u, w - u > + \rho < Tu, w - u > + \rho < Tu, u > - \frac{1}{2} < u, u > . \end{aligned}$$

From which, it follows that the auxiliary variational inequality (42) is equivalent to finding a minimum of $I_3[w]$ on $K \subset H$, where

$$I_3[w] = + \frac{1}{2} < w - u, w - u > - < Tu, w - u >, \tag{45}$$

for $\rho = 1$.

This shows that Fukushima's result is a special case of the auxiliary principle technique with $w = M(u) = P_K[u - Tu]$. Wu, Florian and Marcotte [46] proposed a number of equivalent formulation of the variational inequality problem (15). It has been shown [46] that the solution of the variational inequality (15) can be obtained by solving either one of the following equivalent problems for positive $\alpha$,

(i). $\min_{v\epsilon K} f(v) = \Big\{- < Tv, H(v) - v > -\frac{1}{2} < H(v), v >\Big\}$,

where $H(v)\epsilon\Gamma(v) = arg\ \min_{u\epsilon K} < Tv, u - v > +\frac{1}{2}\ \phi(u,v)$ and $\phi(u,v)\epsilon K \times K \subset H \times H \longrightarrow H$ satisfies some suitable conditions [46].

(ii). $\max_{v\epsilon K}\ \min_{u\epsilon K} \Big\{< Tv, u - v > +\frac{1}{2}\ \phi(u, v >\Big\}$.

(iii). $\min_{v\epsilon K}\ \{Tv, u - v >\}$. $u\epsilon arg\ \min_{v\epsilon K} < T(v), u - v >$.

(iv). $\max_{v\epsilon K}\ \min_{u\epsilon K}\ \{< Tu, u - v >\}$.

It is assumed that the function $\phi(u,v)$ has the property $\phi(u,u) = 0$. We would like to remark that the technique of Wu, Florian and Marcotte [46], cannot be extended to find the equivalence formulation for the mixed type of the variational inequalities considered and studied by Noor [29, 32]. This shows that the auxiliary principle technique developed by Noor [29, 32] is more general and flexible. By a suitable choice of the auxiliary problem, one cannot only suggest a number of equivalent formulations for various types of the variational inequalities, but can also study the existence of the solution. Using the techniques of Fukushima [8], and Wu, Florian and Marcotte [46], one can develop a general descent framework for computing the approximation solution of the general variational inequality (14) and related optimization problems including complementarity problems. As the auxiliary principle technique is related to the minimum problem, so a large number of numerical methods can be used for solving variational inequalities.

## 6. Conclusion And Future Research

In this paper, we have shown that third order obstacle problems arising in physical oceanography can be studied in the general framework of general variational inequalities which was introduced by Noor [23] in 1988. For the recent results regarding sensitivity analysis, iterative method, generalization and application, see Noor [24, 25, 32]. We have used the auxiliary principle to discuss the existence of the solution of variational inequality and have shown this technique can be used to develop an innovative and novel iterative algorithm for nonlinear variational inequalities and related optimization problem. Using the projection technique and its various variant forms numerous iterative algorithms are developed. We have also shown that the auxiliary principle technique is more general and includes many known numerical methods as special cases.

Nevertheless, there are several topics we have not dealt with in this paper that pertain to the general variational inequalities. For example, we have not discussed the regularity of the existence of the solution. The regularity of the solution plays a crucial part in the derivation of the error estimate for the finite element approximation of variational inequalities. The regularity problem for the general variational inequality is mainly an open problem, which is another direction for future research.

We have observed that in a variational inequality formulation, the location of the contact area (moving, free boundary) becomes an integral part of the solution and no special devices are needed to locate it. We have seen that various numerical techniques cannot be applied directly to solve these problems. The use of these methods for solving variational inequalities leads to a complicated search for an approximate solution in comparison with the usual variational method. Using the technique of Lewy and Stampacchia [16], one can characterize the variational inequalities (14) by a sequence of variational equations as

$$< Tu, g(v) > + < \nu(u - \psi)(u - \psi), g(v) >= 0,$$

for all $v \epsilon H$, where $\nu(t)$ is a discontinuous function defined by

$$\nu(t) = \begin{cases} 1, & t \leq 0 \\ 0, & t > 0, \end{cases}$$

is known as the penalty function and $\psi \leq 0$ is defined on the boundary of the domain. Noor and Khalifa [34] used this technique to compute the numerical solution of the obstacle boundary value problem. It would be of interest to compare algorithms studied in this paper to the penalty function method and other standard techniques of variational inequalities.

The theory of general variational inequality does not appear to have developed to an extent that it provides a complete framework for studying various problems arising in physical oceanography, pure and applied sciences. It is true that each of these areas of application require special consideration of peculiarities of the physical problem at hand and the inequalities that model it. Much more research is needed in all these areas to develop a sound basis for application. We have not treated variational inequality for time-dependent problems. In fact, this field has been continuing and will continue to foster new, innovative and novel applications in various branches of physical, pure and applied sciences. The interested reader is advised to explore this field further to discover new and interesting applications.

## References

1. Baiocchi, C., Capelo, A. (1984) *Variational and Quasi Variational Inequalities*, J. Wiley and Sons, London, and New York.

2. Cohen, G. (1988) Auxiliary problem principle extended to variational inequalities, *J. Optim. Theor. Appl.*, **59**, 325-333.

3. Cottle, R.W., Giannessi, F. and Lions, J.L. (1980) *Variational Inequalities and Complementarity Problems: Theory and Applications*, J. Wiley and Sons, London, and New York.

4. Crank, J. (1984) *Free and Moving Boundary Problems*, Clarendon Press, Oxford.

5. Dunbar, S.R. (1993) Geometric analysis of nonlinear boundary value problem from physical oceanography, *SIAM J. Math. Anal.*, **24**, 444- 465.

6. Fichere, G. (1963-64) Problemi elastostatici con vincoli unilaterali: il problema di signorini con ambigue condizione al contorno. *Att Accad. Naz. Lincei Mem. Cl. Sci. Fis. Mat. Natur. Sez. Ia,* (8) **7**, 91-140.

7. Filippov, V.M. (1989) *Variational Principles for Nonpotential Operators*, Vol. 77, American Math. Soc. Providence, Rhode Island.

8. Fukushima, M. (1992) Equivalent differentiable optimization problems and descent methods for asymmetric variational inequality problems, *Math. Programming*, **53**, 99-110.

9. Glowinski, R., Lions, J.L. & Tremolieres, R. (1981), *Numerical Analysis of Variational Inequalities,* North-Holland, Amsterdam.

10. Harker, P.T., Pang, J.S. (1990) Finite dimensional variational inequality and nonlinear complementarity problems: a survey of theory, algorithms and applications, *Math. Programming*, **48**, 161-220.

11. Ierley, G.R., Ruehr, O.G. (1986) Analytical and numerical solutions of a nonlinear boundary-layer problem, *Stud. Appl. Math.*, **75**, 1-36.

12. Isac, G. (1992) *Complementarity Problems*, Lectures Notes in Mathematics, 1528, Springer-Verlag, Berlin.

13. Kikuchi, N., Oden, J.T. (1988) *Contact Problems in Elasticity,* SIAM Publishing Co., Philadelphia, U.S.A.

14. Kinderlehrer, D., Stampacchia, G. (1980) *An Introduction to Variational Inequalities and Their Applications*, Academic Press, London.

15. Korpelevich, G.M. (1976) The extragradient method for finding saddle points and other problems, *Ekon. I. Math. Meody*, **12**, 747-756.

16. Lewy, H., Stampacchia, G. (1969) On the regularity of the solutions of the variational inequalities, *Comm. Pure Appl. Math.*, **22**, 153-188.

17. Lions, J.L., Stampacchia, G. (1967) Variational inequalities, *Pure Appl. Math.*, **20**, 493-519.

18. Luo, Z.O., Tseng, P. (1992) Error bounds and convergence analysis of feasible decent methods: a general approach, Preprint.

19. Mosco, U. (1976) *Implicit variational problems and quasi variational inequalities*, Lect. Notes Math. 543, Springer-Verlag, Berlin, 83- 126.

20. Murty, K. (1988) *Linear Complementarity, Linear and Nonlinear Programming,* Heldermann Verlag, Berlin, 1988.

21. Noor, M.A. (1985) An iterative scheme for a class of quasi variational inequalities, *J. Math. Anal. Appl.*, **110**, 463-468.

22. Noor, M.A. (1987) General nonlinear variational inequalities, *J. Math. Anal. Appl.*, **126**, 78-84.

23. Noor, M.A. (1988) General variational inequalities, *Appl. Math. Letters*, **1**, 119-122.

24. Noor, M.A. (1992) General algorithm and sensitivity analysis for variational inequalities, *J. Appl. Math. and Stoch. Anal.*, **5**, 29- 42.

25. Noor, M.A. (1993) Some recent advances in variational inequalities, Preprint.

26. Noor, M.A. (1992) An iterative algorithm for nonlinear variational inequalities, *Appl. Math. Letters*, **5**(4), 11-14.

27. Noor, M.A. (1992) Iterative algorithms for quasi variational inequalities, *PanAmer. Math. J.*, **2**(2), 17-26.

28. Noor, M.A. (1993) Iterative methods and sensitivity analysis for general quasi variational inequalities, *J. Nat. Geometry*, **3**, 39-58.

29. Noor, M.A. (1993) General algorithms for variational inequalities I, *Math. Japan,* **38**, 47-53.

30. Noor, M.A. (1992) General algorithm for variational inequalities, *J. Optim. Theory Appl.*, **73**, 409-413.

31. Noor, M.A. (1993) Wiener-Hopf equations and variational inequalities, *J. Optim. Theory Appl.*, **73**, in press.

32. Noor, M.A. (1994) General auxiliary principle for variational inequalities, *PanAmercan Math. J.*, **4**(1), in press.

33. Noor, M.A., Noor, K.I., Rassias, Th.M. (1993) Some aspects of variational inequalities, *J. Comput. Appl. Math.*, **47**, 285-312.

34. Noor, M.A., Khalifa, A.K. (1987) Cubic splines collocation methods for unilateral problems, *Internat. J. Engng. Sci.*, **25**, 1527-1530.

35. Oden, J.T., Kikuchi, N. (1980) Theory of variational inequalities with applications to problems of flow through porous media, *Internat. J. Engng. Sci.*, **18**, 1173-1284.

36. Panagiotopoulos, P.D. (1985) *Inequality Problems in Mechanics and Applications,* Birkhauser, Boston.

37. Petryshyn, W.V. (1962) Direct and iterative methods for the solution of linear operator equations in Hilbert space, *Trans. Amer. Math. Soc.*, **105**, 136-175.

38. Petryshyn, W.V. (1992) *Approximation, Solvability of Nonlinear Functional and Differential Equations,* Marcel Dekker, New York.

39. Pitonyak, A., Shi, P., Shillor, M. (1990) On an iterative method for variational inequalities, *Numer. Math.*, **58**, 231-242.

40. Rodrigues, J.F. (1987) *Obstacle Problems in Mathematical Physics,* North-Holland, Amsterdam.

41. Shi, P. (1991) Equivalence of variational inequalities with Wiener-Hopf equations, Proc. Amer. Math. Soc., **111**, 339-346.

42. Shi, P. (1990) An iterative method for obstacle problems via Green's functions, *Nonlinear Anal.*, **15**, 339-344.

43. Speck, P.O. (1985) *General Wiener-Hopf Factorization Methods,* Pitman Advanced Publishing Program, London.

44. Stampacchia, G. (1964) Formes bilineaires coercitives sur les ensembles convexes, *C.R. Acad. Sci. Pari*, **258**, 4413-4416.

45. Tonti, E. (1984) Variational formulation for every nonlinear problem, *Internat. J. Engng. Sci.*, **22**, 1343-1371.

46. Wu, J.H., Florian, M., Marcotte, P. (1993) A general descent framework for the monotone variational inequality problem, *Math. Programming*, **61**, 281-300.